How To Draw Dinosaurs

Volume 3

By Tracy L. Ford

ISBN-13: 978-1717484703

ISBN-10: 1717484700

Table of Contents

Table of contents……………………………………………………………………...2
Dedication……………………………………………………………………………..3
Acknowledgements…………………………………………………………………. 4
Chapter 1). Spiky Pachycephalosaurs?…..…………………………………………...5
Chapter 2). The End of Ornithopods………………………………………………....9
Chapter 3). The whole scene……………………………………………………….12
Chapter 4). Stegosaurs: Plates, Splates and Spikes, Part 1………………………….17
Chapter 5). Stegosaurs: Plates, Splates and Spikes, Part 2………………………….22
Chapter 6). Con- or Pro-boscis Diplodocids or Trunk-aided diplodocids, Science Fact or Fiction?..28
Chapter 7). Ammonites are not Nautiloids, part 1…………………………………..33
Chapter 8). Ammonites are not Nautiloids, part 2 (and other fossil cephalopods)……..39
Chapter 9). A little cartilage goes a long way (Back to dinosaurs…)……………….44
Chapter 10). Dinosaurs, can you dig them? Fossorial fossils…………………….49
Chapter 11). Neck muscles and feeding strategies in large theropods……………55
Chapter 12). And now, Back to Sauropods…………………………………………63
Chapter 13). Hadrosaurs revisited…………………………………………………...70
Chapter 14). Swimming with dinosaurs, part 1……………………………………..75
Chapter 15). Swimming with dinosaurs, part 2……………………………………..84
Chapter 16). Were Stegosaurs bipedal?... 90
Chapter 17) Face Fighting Ceratopians……………………………………………...100
Chapter 18). The dinosaurs with a bad name…Oviraptorids, Part 1……………….106
Chapter 19). The dinosaurs with a bad name…Oviraptorids, part 2……………….112
Chapter 20). Bulking up Sauropods (and dinosaurs in general)…………………….117
Chapter 21). Dull or colorfully? What did dinosaurs look like?...................................124
Chapter 22). Aquatic Pittacosaurs (finally) Part one………………………………130
Chapter 23). Aquatic Pittacosaurs (finally) Part two……………………………….136
Chapter 24). To know the nose, Part 3 (Sauropods)………………………………...145
Chapter 25). And Now, The End Of Dinosaurs……………………………………...150

Dedication

I dedicate this book to my good friend and long-time supporter, Tom Kaye. We have had many good conversations, he has given me ideas, and he suggested the name for my website, Paleofile. His research spans paleontology (his research on laser florescent light has shown remarkable fossil preservation), astronomy, and other interesting ventures. I look forward too many more years of his support and help in my research.

Acknowledgements

I would like to thank Mike Fredericks for letting me write the "How to Draw Dinosaurs" articles, and for his editing. In late 1995 (I think) I started buying Mike's magazine "Prehistoric Times" and bought the back issues at the time to fill out my collection (I'm happy to say I have a complete collection of Prehistoric Times). At the second Dinofest held in Arizona (1996) I had the pleasure of dining with Mike. It was then that I brought up the idea of starting a series of articles on how to better depict dinosaurs by using their anatomy. He was up for it and shortly after that I started to write the articles for each issue of the magazine. I have fun writing the articles and I'm glad to hear from many people who use them. Luckily, I still have lots of things to write about so the series will continue for quite some time. I would also like to thank my family for their support of my passion for dinosaurs, George Olshevsky, whom has been my paleontological friend and mentor, Darren Tanke who, along with George, has helped me in getting my illustrations published, the paleontologists who have helped me over the years: Jim Kirkland, and Ken Carpenter (who both helped me become a published 'paleontologist'), along with Tom Demere at the San Diego Natural History Museum, Bob McCord, Debbie Boaz and all the rest at the Mesa Southwest Museum (now called the Arizona Museum of Natural History), Dan Chure, Ralph Molnar, Peter Galton, Phil Currie, Jack Horner, John McIntosh, Spencer Lucas, Mike Brett-Surman, Don Glut, Greg Paul, Stephen Czerkas, Mark Hallett, and I apologize to all the others that I haven't named.

Ford, T. L., 2005, How to Draw Dinosaurs. Spiky Pachycephalosaurs? Prehistoric Times, n. 73, p. 20.

Chapter 1

Spiky Pachycephalosaurs?

The first skulls found of *Pachycephalosaurus* have rounded knobs on the posterior part of the skull. This has been the rule for their reconstructions. Approximately 40 years later *Stygimoloch* was described. *Stygimoloch* has a narrow dome, but very long horizontal spikes on the back of the skull. More than a decade ago Mike Triebold found a partial skull and skeleton of a pachycephalosaur. This animal has spikes and not knobs. I remember seeing it for the first time at my first Tucson Rock/Mineral/Fossil show. I talked to Mike about it and I said I thought it was just a *Stygimoloch* with shorter spikes. He was unconvinced. Strangely this new find was missing the dome, but Mike said he did find 3 different domes, one like *Stygimoloch* and the others like *Pachycephalosaurus*. The Black Hills Institute has recently found a *Pachycephalosaurus* skull with spikes not knobs. While not as long as Mike's specimen, this does show it had spikes not knobs. So, what about the type specimens. I now believe the spikes had been worn down to knobs and in life they should have been spikes. The front of *Pachycephalosaurus grangeri* have spikes not knobs and this is probably the same with the back of the skull.

Bibliography

Brown, B. B., and Schlaikjer, E. M., 1943, A study of the Troodont Dinosaurs with the description of a new genus and four new species: Bulletin of the American Museum of Natural History, v. 82, article 5, p. 120-149.

Galton, P. M., and Sues, H-D., 1983, New data on pachycephalosaurid dinosaurs (Reptilia: Ornithischia) from North America: Canadian Journal of Earth Sciences, v. 20, p. 462-472.

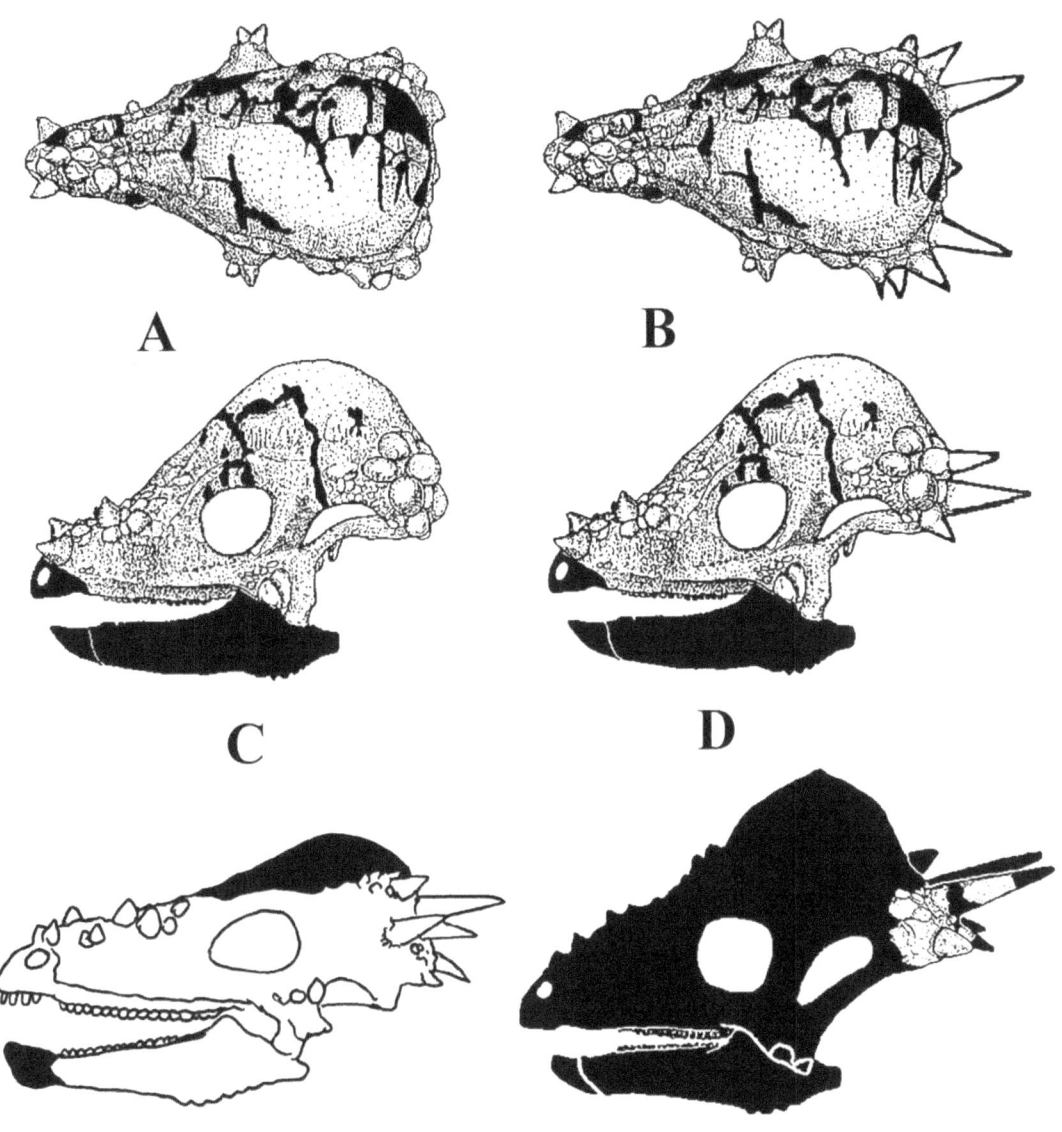

Figure 1). Pachycephalosaur skulls. A) Type of *Pachycephalosaurus grangeri* showing knobs in dorsal and lateral views; B) *Pachycephalosaurus* showing spikes, in dorsal, and lateral views; C) Sandy Site Pachycephalosaur; D) *Stygimoloch* skull.

Small eared Woolly's?

Recently a Japanese Scientific team has examined a new frozen Woolly Mammoth from the Siberian tundra. The Woolly Mammoth is 18,000 years old and was unveiled at the 2005 World Exposition in Aichi, Japan. The team showed that the DNA is closer to the Indian Elephant than the African. This is not really new news, but what is new is the ridiculously small ears! The ears are tiny. In 1977 two baby mammoths were found. They were 6 to 12 months old and it has small ears. The famous Berezovka Mammoth had the skull and some soft tissue, but no ears were left. The diorama of the Berezovka Mammoth is a reconstruction. But small ears make sense. Elephants use their ears to cool themselves off. In a frozen environment they wouldn't need big ears. So, every picture, sculpture etc. of a large eared Woolly is wrong. Who knew?

Figure 2) Woolly Mammoth with A) correct small ears; B) incorrect large ears.

PreHistoric Times

#74 Oct/Nov 2005

The Latest News on Feathered Dinosaurs

A PT Pictorial: Feathered Dinosaurs

Front cover art by Fabio Pastori

Everything you wanted to know about Tyrannosaurus rex
The PT Interview
Dr Gregory Erickson

Kong
King of Skull Island
Part 2

Chapter 2

The End of Ornithopods

One of the things that struck me when I started to draw dinosaur skeletons was how differently mine came out looking from ones that I'd seen before. I remember drawing a *Hypsilophodon foxii* skeleton and noticing the pelvic area looked different. The pubis and ischium were lower than I thought and extended longer in relation to the caudal vertebrae. The pubis/ischium extended beyond the knee. But this area varies from ornithopod to ornithopod. In *Thescelosaurus* the pubis/ischium extends a little beyond the ilium and less so in *Camptosaurus*. In iguanodontids the pubis/ischium extends to just past the 4th caudal vertebrae. In hadrosaurs this goes to the extreme with it extending to the 7th to 8th caudal vertebrae.

The pubis/ischium of ornithopods was solid and did not move, but when looking down from the dorsal area you would not see this area because the caudal vertebrae covers it. The ischium had tail muscles attached to it and the caudal vertebrae. The larger the ischium the larger the muscle area. The neural spines in iguanodontids and hadrosaurids had ossified tendons These tendons for a lattice work and helped stiffen the tail. They couldn't move the tail much at all. Contrary to popular belief, there is now way they could use their tail to swim.

Not to be curd but what about the anal opening? How did it look? Well, since there are no living dinosaurs we wouldn't' know for sure. It could have either been a slit vertical or horizontal. It's funny; if you buy all the Battat toy dinosaurs you'll notice that there are an equal amount of both. This is because sculptor Greg Wenzel did it one way, and the sculptor Dan Lorusso did it the other!

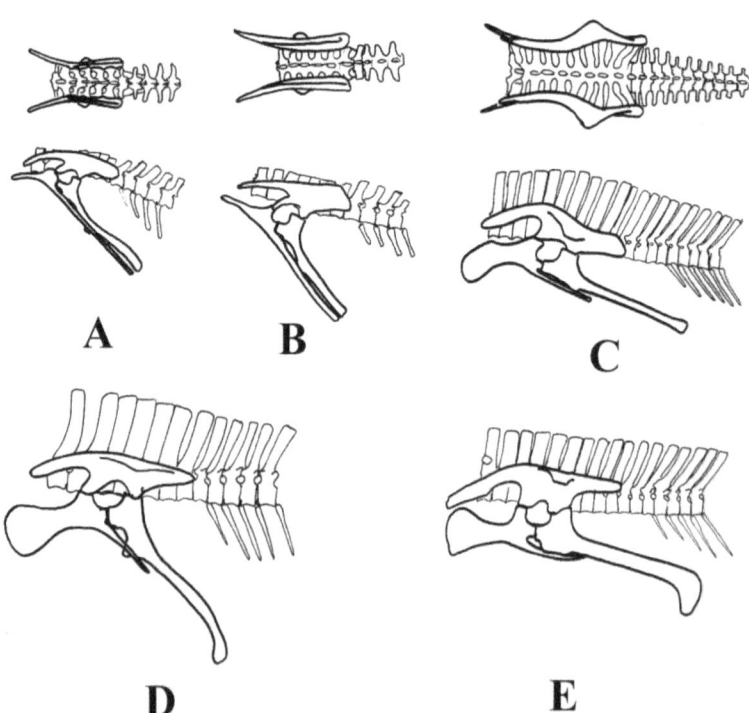

Figure 1) Pelvic region of some ornithopods; A) lateral and dorsal view of *Hypsilophodon* (after Paul); B) lateral and dorsal view of *Thescelosaurus*; C) Lateral and dorsal view of '*Kritosaurus*' *incurvimanus* (after Paul); D) *Iguanodon*: and E) *Corythosaurus*.

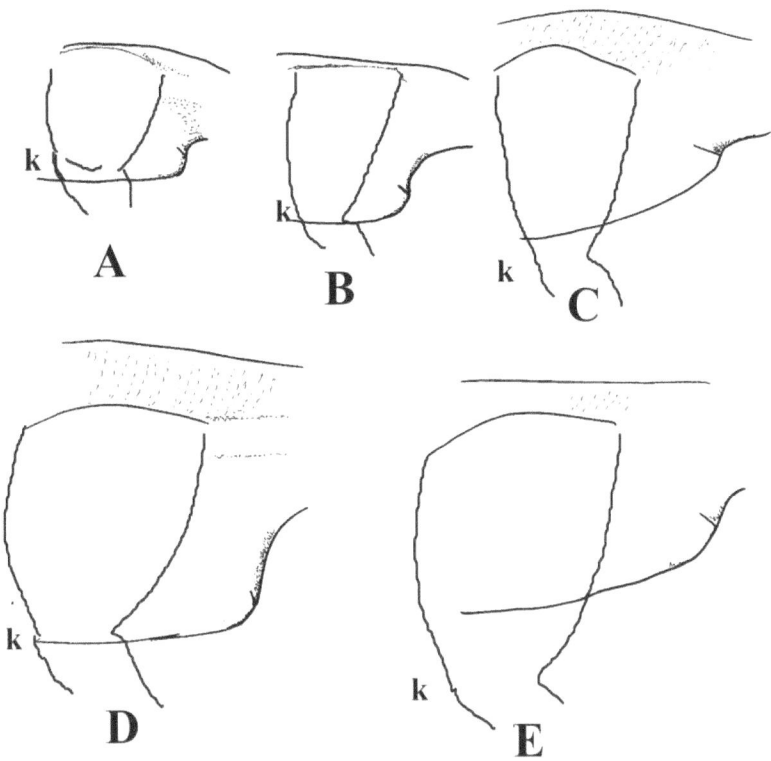

Figure 2) Pelvic region of some ornithopods showing the position of the (k) knee: A) *Hypsilphodon*; B) *Thescelosaurus*; C) '*Kritosaurus' incurvimanus*; D) *Iguanodon*; E) *Corhtyosaurus*.

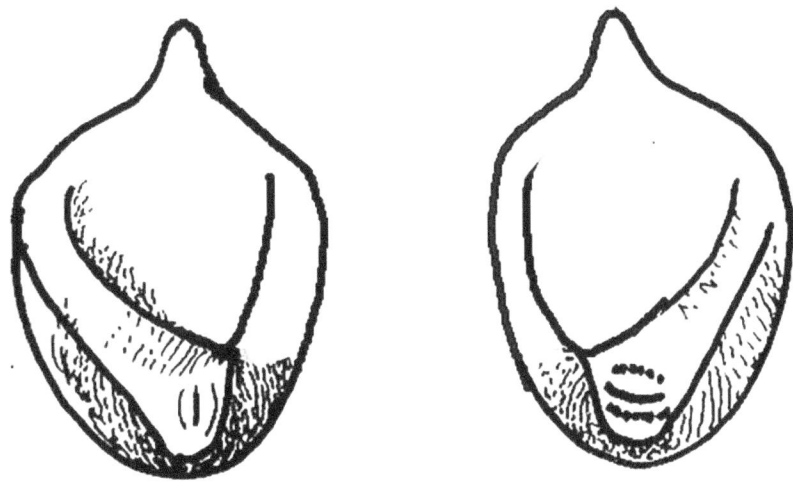

Figure 3) Posterior end of an ornithopod showing the possible anal opening.

11

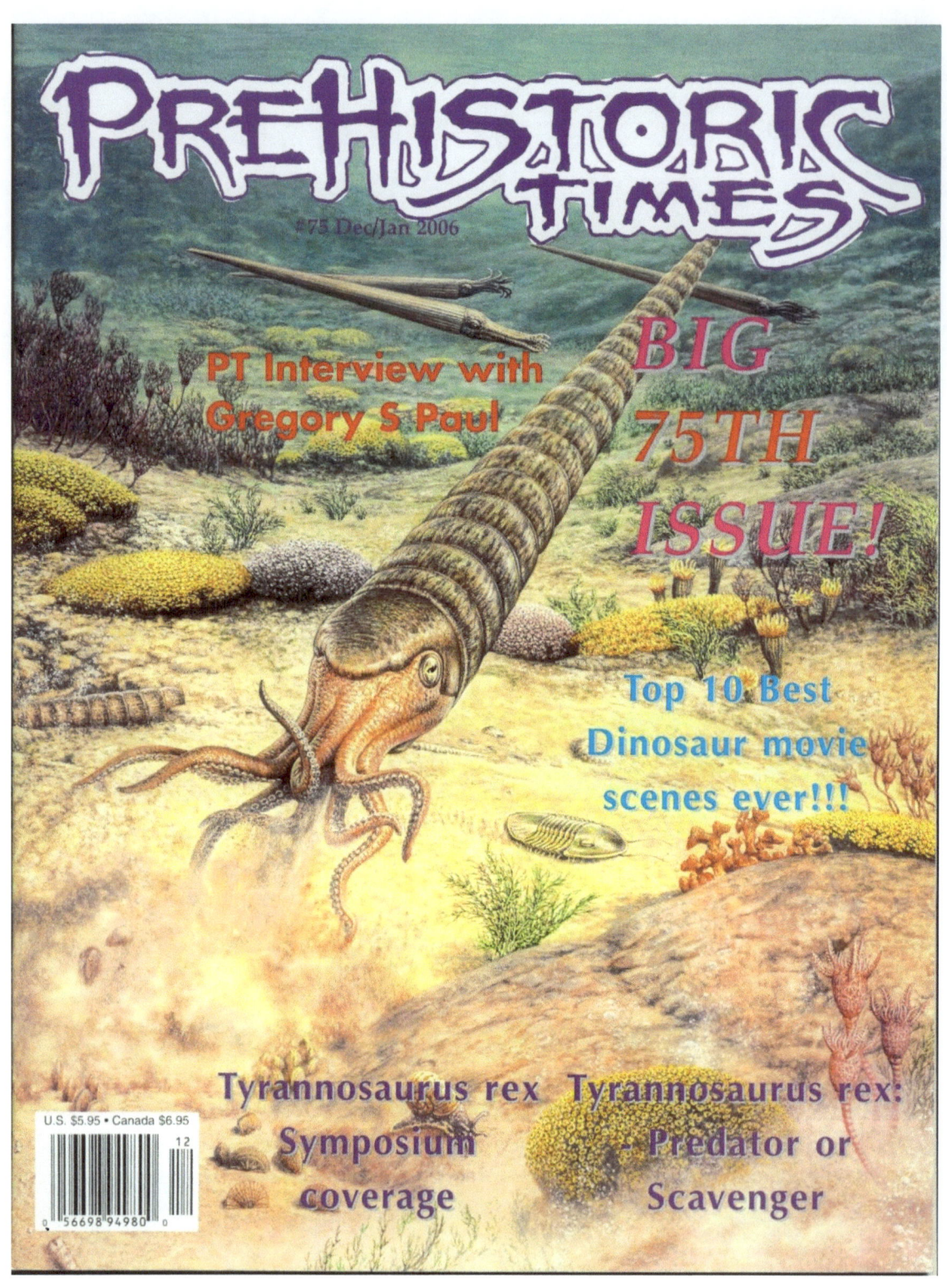

Ford, T. L., 2006, How to Draw Dinosaurs. The whole scene…: Prehistoric Times, n. 75, p. 20-21.

Chapter 3

The whole scene…

This time around is going to be different. No skeletal elements, no footprints, feathers, scales or horns etc. The idea for this article came about from a friend of mine who suggested I should include birds in my Late Cretaceous scenes. That got me thinking about other animals and not just dinosaurs.

As Dr. Ken Carpenter said at the Burpee Tyrannosaur Symposium, dinosaurs didn't live in parking lots; i.e. put in vegetation and other animals.

So, here are some ideas…

1When doing an ocean scene, put in fish, sharks, rays, chimera, squids, ammonites nautiloids, jellyfish etc. I though jellyfish would look great in a picture so I illustrated a tylosaur swimming in a school of jellyfish (Figure 1). If portraying an ocean bottom include crabs, lobsters, clams, oysters, coral, sea cumbers, starfish, brittle stares, etc. or equivalent animals living in a modern environment.

On the seashore add horseshoe crabs (check out which kind), land crabs, vegetation, etc.

When drawing a terrestrial scene, check out similar modern environments to see what kind of animals live there today (invertebrates, plants) and do some research for any information regarding what lived then. This is where paleofile will come in handy You can check my locality lists to see what lived with what and when (http://www.paleofile.com).

Use the same 'kind' of color patterns of modern animals, but don't put a green iguana in a desert! Perhaps go to a local county fair and take pictures of all the different color patterns animals have. Spiders would be cool. Draw some spider webs, dragonflies, flies, bugs of some sort, and eve termite mounds (Figure 2). As mentioned, show birds if they were around during the age you are working on. In the distance make the silhouette 'V' for birds flying, or pterosaurs. Also, don't forget about turtles, lizards, snakes, crocodiles, sphenodonts…

Plants are hard for me. I don't have a lot of data on them, and usually only leaf or partial leaf/leaves are preserved so sometimes you really don't know what the plant or tree would look like. See out references: a good book that I ran across when I was younger is great for plants. It is titled: *All New Dinosaurs and their friends from the Great Recent Discoveries*. Long, Robert A., and Samuel P. Welles, 1988. Bellerophon Books, 48pp.

It's still being published today, and it is great for showing what kind of plants were around during the Mesozoic. There was another one just for the Late Cretaceous. I wish they did more of them.

Check out any Doug Henderson painting. He is a master of prehistoric scenes, as is Mark Hallet, Greg Paul, Bill Stout, etc. Also get your hands on the book: *Dinosaurs a Global view*. Czerkas, Sylvia, J., and Stephen A. Czerkas, 1990. Dragon's World: 247pp.

Place your dinosaurs behind vegetation or passing through it.

Try and put in some shadows. Determine the time of day, where the sun would be, and how the shadows would be cast. Put in different kinds of clouds, different kinds of weather, rain, wind, etc.

Don't use just side (lateral) views; try different angles, or different behaviors (Figure 3). Use your imagination and have some fun.

Figure 1) Tylosaur swimming amongst jellyfish.

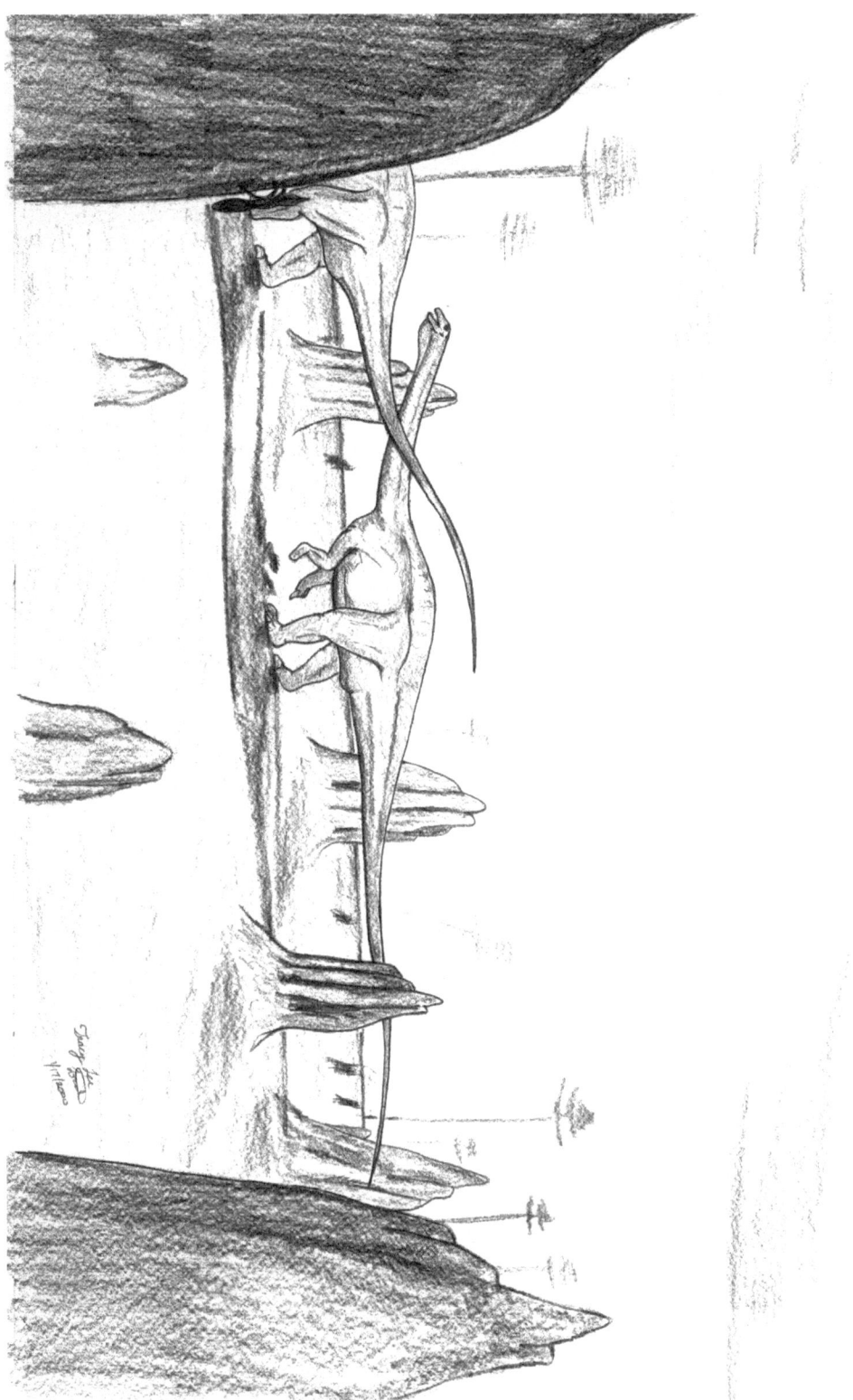

Figure 2) Apatosaurs walking bipedally in a termite mound field.

Figure 3) *Velociraptor* running down a dune.

Ford, T. L., 2006, How to Draw Dinosaurs. Stegosaurs: Plates, Splates and Spikes, Part 1. Prehistoric Times, n. 76, p. 20-21.

Chapter 4

Stegosaurs: Plates, Splates and Spikes, Part 1

Stegosaurus is one of the most famous and easily recognized dinosaurs from North America. It has a certain draw to people. Perhaps it's the large plates and spikes on its tail, or its small head and brain? The second article I wrote for PT was on the orientation of the tail spikes of Stegosaurs/us, but I've never written about the plates of stegosaurs. The first part of my tow part article will be on stegosaurs in general and the second on *Stegosaurus* specifically. Stegosaurs are known from North America, Asia, Europe, and now South America (known from a few skeletal elements).

The plates of stegosaurs vary not only in shape but also in size. The bases of the plates differ; in the cervical to mid dorsal the bases are wide, while from the mid dorsal to the caudals the bases are thin (Figure 1). The wider bases where held more loosely than the thinner ones. The thinner plates were more deeply imbedded in the skin. The plates are arranged in a double parallel row down the center of the back, while in *Stegosaurus* the plates from the middle of the back to the last one of the tail was offset from each other though still in a double row (more in that next time) (Figure 2). A double row of osteoderms is the primitive condition in tetrapods. The plates can be square, leaf-shape, round, spike-like, and even round with a spike (a *splate* so named by George Olshevsky) (Figure 3). Plate size and overall animal size do not correlate to each other. *Dacentrurus* was as large as *Stegosaurs*, but it's plates are about 1/3 or so as large, even though they lived at the same time (Figure 4). The purpose of the plates is highly controversial. Sexual selection, individual variation or for heat transfer all have been suggested. I favor the first two.

Some stegosaurs had a specialized shoulder spike. Originally this spike was thought to have been on the pelvis. *Kentrosaurus* (from East Africa) was often shown with the spikes extending from its hip (Figure 4). It wasn't until much later that it was determined that the spikes actually were from the shoulder region. *Huayangosaurus, Lexovisaurus, Dacentrurus, Yingshanosaurus* all have been found with these spikes. These spikes are large and face upward and backward. *Tuojiangosaurus* was believed to have had a shoulder spike, but that specimen (nearly complete and articulated) was determined to actually belong to a new genus, *Gigantspinosaurus sichuanensis* Ouyang, 1992. This stegosaur is relatively unknown to paleontologist. The paper was very short, and unknown. There is a new Japanese traveling exhibit that has a mounted skeleton of *Gigantspinosaurus* and the pose it is in makes it look like a monster from a Japanese movie. The plates are very small, but the shoulder spines are HUGE !!! Larger than any other stegosaur. As far as I know this will be the second time *Gigantspinosaurus* skeleton has seen print (Figure 5). The first one is from a 1992 Japanese book which at that time referred it to *Tuojiangosaurus*. The mounted skeleton has the spike facing down and backwards, though I believe looking at the skeleton in situ the spike should be facing up and back (fig).

Bibliography

Dong, Z.-M., 1992. Dinosaurian Faunas of China. China Ocean Press: 188pp.

Dong, Z.-M., Li, X., Zhou, S., and Zhang, Y., 1977, On the Stegosaurian remain from Zigong (Tzekung), Sechuan Province: Vertebrata PalAsiatica, v. 15, n. 4, p. 307-302.

Dong, Z.-M., Tang, Z., and Zhou, S., 1982, Note on the new Mid-Jurassic stegosaur from Sichuan Basin, China: Vertebrata PalAsiatica, v. 10, n. 1, p. 83-87.

Dong, Z.-M., Zhou, S., and Zhang, Y., 1983, The Dinosaurian remains from Sichua6n Basin, China: Palaeontologica Sincia, Whole Number 162, new series C, n. 23, p. 1-145.

Galton, P. M., 1991, Postcranial remains of stegosaurian dinosaur *Dacenterurus* from Upper Jurassic of France and Portugal: Geologica et Palaeontologica, v. 25, p. 299-327.

Galton, P. M., Brun, R., and Rioult, M., 1980, Skeleton of the Stegosaurian dinosaur *Lexovisaurus* from the Lower part of the Middle Callovian (Middle Jurassic) of Argences (Calvados), Normandy: Bulletin de la Societe geologique de Normandie, Anis Museum de Have, tomo 67, fasc 4, p. 39-60.

Gilmore, C. W., 1914, Osteology of the armored dinosauria in the U. S. National Museum with special reference to the genus *Stegosaurus*: Bulletin of the United States National Museum, v. 89, p.1-140.

Henning, E., 1915, *Kentrosaruus aethiopicus*, der Stegosauridae des Tendaguru: Sitzungsberichte Naturforschender Freunde zu Berlin, 1915, n. 6, p. 219-247,

Henning, E., 1916, Zweite Mittelung uber den Stegosauriden vom Tendaguru Sitzungsberichte Naturforschender Freunde zu Berlin, 1915, p. 175-182.

Henning, E., 1925, *Kentruruosaurus aethiopicus*, die Stegosaurier- Funde von Tendaguru, Deutsch-Ostafrika: Palaeontologrpahica Supplement v. 7, n. 1.1, p. 101-254.

Quyang, H., 1992, Discovery of *Gigantspinosaurus sichuanensis* and its scapular spine orientation: The Satellite Meeting of the First Youth Academic Annual Confereces by Chinese Science Association, Abstracts and Summaries for Youth Academic Symposium on New Discoveries and Ideas in Stratigraphic Paleontology, Nanjing, Dec. 1992, 3pp? (translated by Jimin Yu, University of Cambridge, October, 2004).

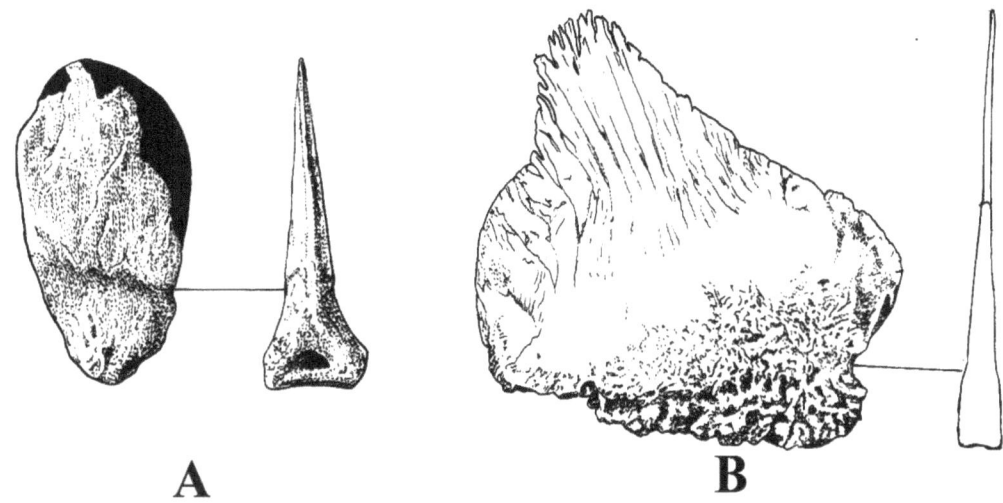

Figure 1). Plates of *Stegosaurus*. A) A cervical plate showing the wide bottom; B) Pelvic plate showing a narrow bottom. The line indicates where the body line was.

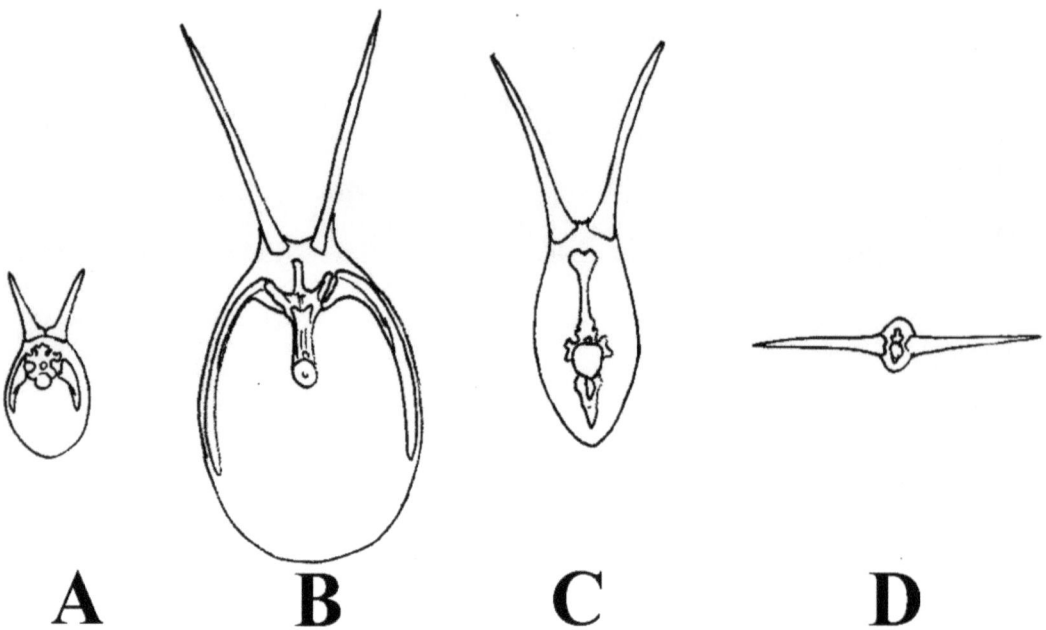

Figure 2): Cross section of *Stegosaurus* showing how the plates were imbedded in the body; A) Mid-cervical region; B) Mid-dorsal region; C) Mid-Caudal region; D) tip of tail.

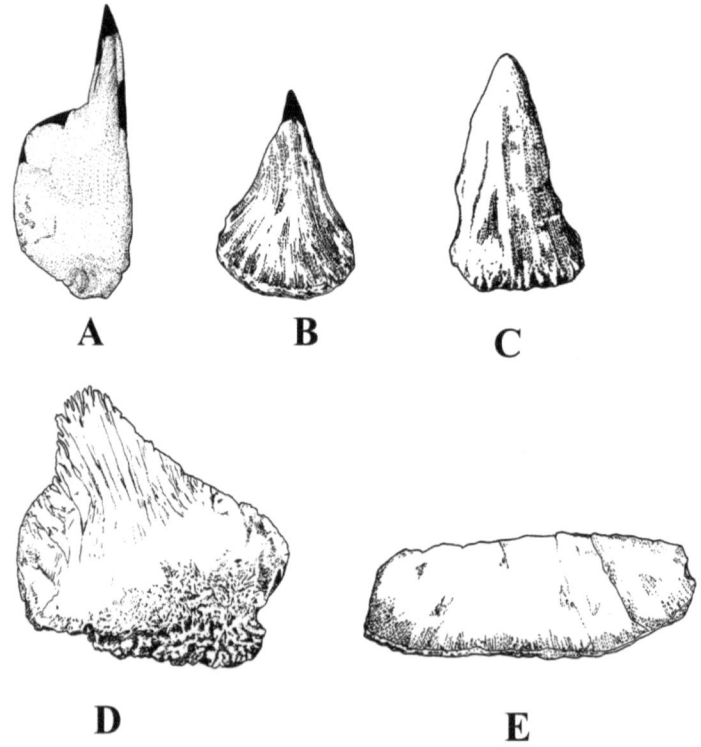

Figure 3): Plates of stegosaurs; A) Splate of *Lexovisaurus*; B) Dorsal plate of *Chungkingosaurus* ; C) Dorsal plate of *Tuojiangosaurus*; D) Dorsal plate of *Wuerhosaurus*; E) Pelvic plate of *Stegosaurus*.

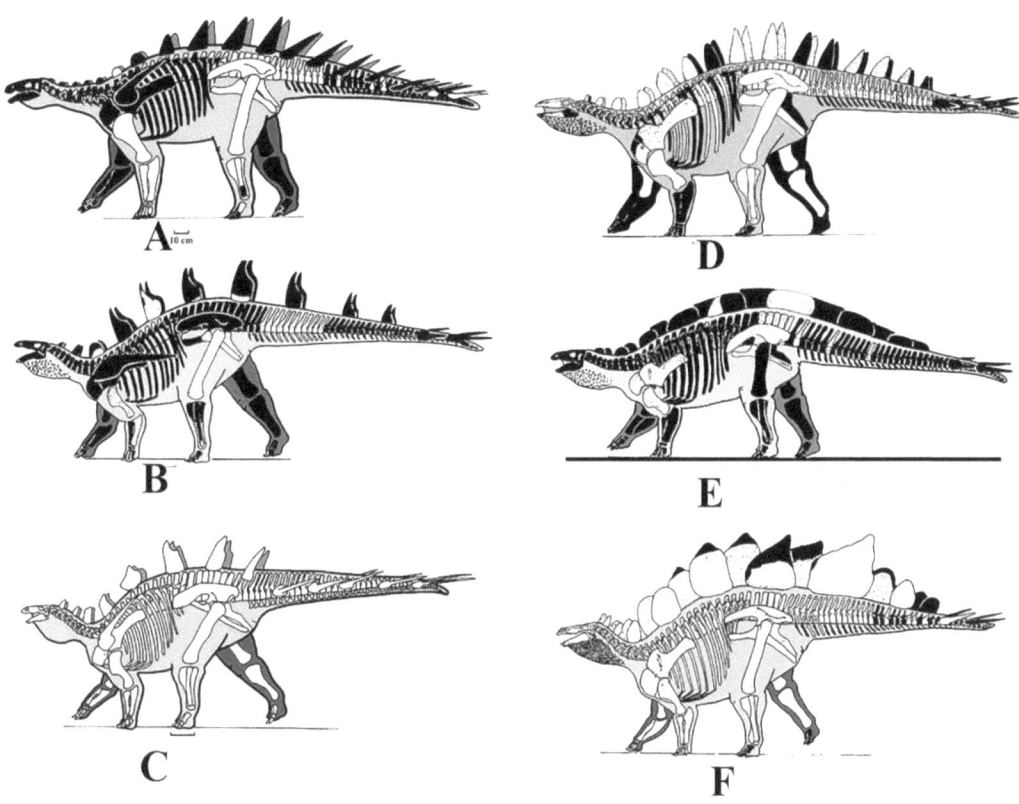

Figure 4): Stegosaur skeletons; A) *Dacentrurus*; B) *Lexovisaurus*; C) *Kentrosaurus*; D) *Tuojiangosaurus*; E) *Wuerhosaurus*; F) *Stegosaurus*.

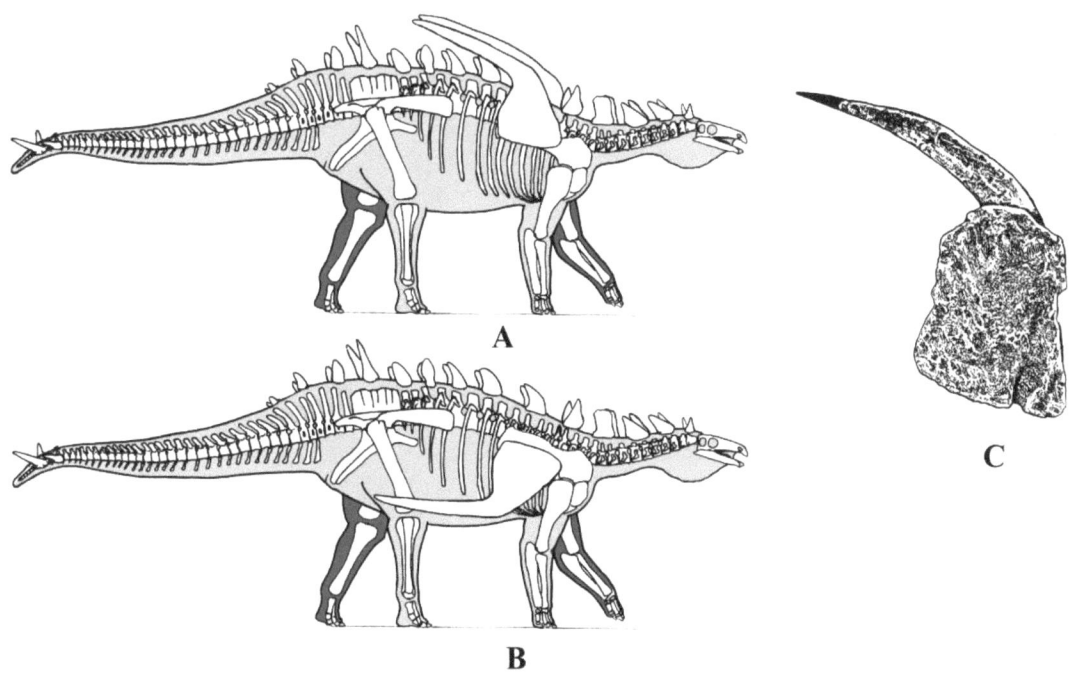

Figure 5): Skeleton of *Gigantospinosaurus sichuanensis*; A) A corrected version of the shoulder spike; B) Illustrated from the mounted skeleton; C) Ventral view of shoulder spine.

Ford, T. L., 2006, How to Draw Dinosaurs. Stegosaurs: Plates, Splates and Spikes, Part 1. Prehistoric Times, n. 76, p. 20-21.

Chapter 5

Stegosaurs: Plates, Splates and Spikes, Part 2

Stegosaurus is by far the most famous and most popular armored dinosaur and *Stegosaurus stenops* is the most recognizable species. There are several species of *Stegosaurus*. Depending on who you follow there are either 3 to 5 species. The type species of *Stegosaurus* (*S. armatus*) has never been fully described. It is a partial skull and skeleton. The problem is the matrix is very hard and is like concrete. Marsh found other more easily prepared specimens, so it was left on the shelves at Yale. I think it's being prepared now though. In the new 2nd edition of Dinosauria Galton and Upchurch sank *S. ungulatus*, *S. duplex and S. sulcatus* into *Stegosaurus armatus*. I asked Galton about this and he's not sure if that is correct. So, for now I'm keeping them separate, except for *S. duplex* being sunk into *S. ungulatus* (which everyone believes).

The most abundant species of *Stegosaurus* is *S. stenops*, with *S. ungulatus* being the second abundant. *S. sulcatus* is known from a partial skeleton with a plate and a few interesting tail spikes (though Bakker mentions a few skeletons that haven't been described). *S. longispinus* is known from a fragmentary skeleton, a few plates and spikes (Editor's note: *Stegosaurus longispinus* has been renamed *Alcovasaurus* by Galton, & Carpenter, 2016). There maybe another *Stegosaurus* species at Dinosaur National Monument. The specimen on 'cliff wall' has very pointed plates and at one-time Bakker wanted to name it a new species. I've asked some paleontologist about these plates and to find out if they are real. I was told that the 'preparators' took off too much 'bone', but for the life of me I can't 'fit' them into either *S. ungulatus or S. stenops* plate. I believe they are the correct shape. At one-time Bob Bakker wanted to make this a new species. (Figure 1).

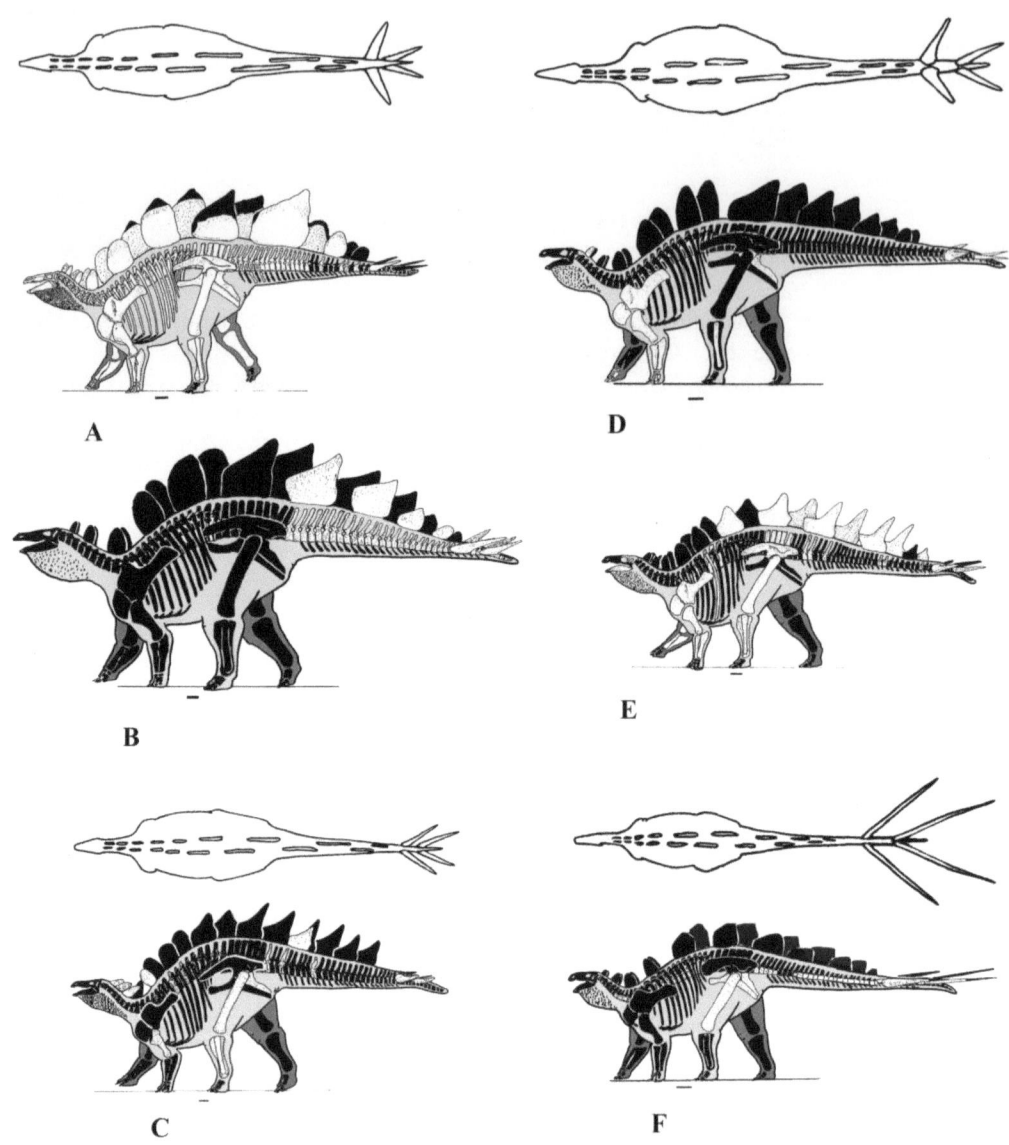

Figure 1). Different species of *Stegosaurus* drawn to the same scale. A) *Stegosaurus stenops* (USNM 4934), dorsal and lateral view; B) *Stegosaurus stenops* USNM 4714; C) *Stegosaurus ungulatus* in dorsal and lateral view; D) *Stegosaurus sulcatus* in dorsal and lateral view; E) Stegosaur at Dinosaur National Monument; F) *Alcovasaurus* (*Stegosaurus*) *longispinus* in dorsal and lateral view.

The Plates

The plates are known from each area (cervical, dorsal, pelvic and caudal) in *S. stenops* and *S. ungulatus*. The other specimens have plates associated with them, but not a complete series. The cervical and dorsal series in *S. ungulatus* are angled backward, while in *S. stenops* they are more vertical. The posterior dorsal and caudal plates in *S. stenops* have a pointed apex on the upper posterior edge, they are very broad, are larger than other species and fuller. In *S. ungulatus* the plates are smaller, the apex is lower and more pointed while in the DNM specimen the apex point is extremely pointed.

The size of the plates doesn't indicate the size of the animal. USNM 4934 (the road kill) is a nearly complete specimen and is smaller (skeletally) than USNM 4714, yet 4934 has larger plates than 4714.

One of the prevailing theories about the plates is that they were used for thermoregulation. But it is only the plates of *S. stenops* that have been used in that theory and not the other thinner plated species. Since the Morrison was warmer the animal wasn't trying to get warm, but to shed heat. They claim blood would flow through the plate and be cooled by the wind. The plates are porous but not in such a way that would allow blood to flow in and out of the plate. The porosity was due to growth. The plates would also have a keratinous covering or horny sheath. This would increase the size to about 10 to 20%. (Figure 2).

Then what were the plates used for? Species identification, sexual dimorphism, and individual variation, as has been recently been reported on.

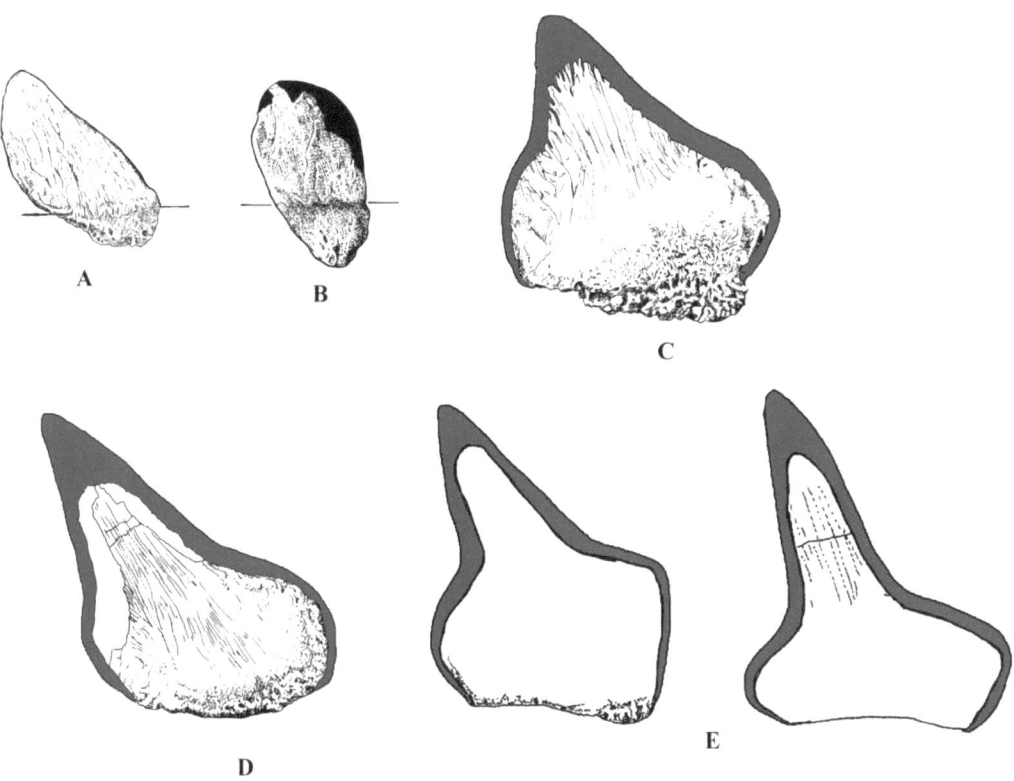

Figure 2). Plates in *Stegosaurus*; A) Cervical plate of *Stegosaurus ungulatus*; B) Cervical plate of *Stegosaurus stenops*; C) Dorsal plate of *Stegosaurus stenops*; D) Dorsal plate of *Stegosaurus ungulatus*; E) two dorsal plates of the Stegosaur at Dinosaur National Monument.

Tail spikes

The tail spikes also vary in the different species of *Stegosaurus*. I won't go into detail about the position of the spikes since I've written about it already in PT (way back in 1997, which was my second article for PT, also chapter two in my How to Draw Dinosaurs volume 1). *S. armatus* was believed to have had 4 sets of tail spikes, but this has recently been shown that that is not the case, it had the typical 2 sets of spikes. *S. stenops* has a broader based first caudal spike that sat more perpendicular to the tail, while the thinner second pair was smaller and sat more angled to the tail. In *S. ungulatus* the spikes base is shallower and more beveled and are both more like the smaller second pair of *S. stenops*. The first caudal spikes of *S. sulcatus* have a very broad base while the last pair are more typical in shape. Because the base is so large it must have been placed higher on the tail than the other species. *Alcovasaurus* (*Stegosaurus*) *longispinus* has a very long spikes with a shallow base. It has been speculated that this species might be a kentrosaur. A few years at the Tucson Rock/Fossil show I bought a cast of an extremely long stegosaur tail spike. It has the length of *Alcovasaurus* (*Stegosaurus*) *longispinus*, but the base like *S. sulcatus*. No other information was available to me about where it was found, what age, etc. The tips of many of the spikes are broken off and were longer in life (Figure 3). Also, like the plates, they would have had a keratinous covering and would have been about 10 to 20% larger.

When illustrating *Stegosaurus stenops* vary the size of the plates and use *S. ungulatus* from time to time. Both lived at the same time.

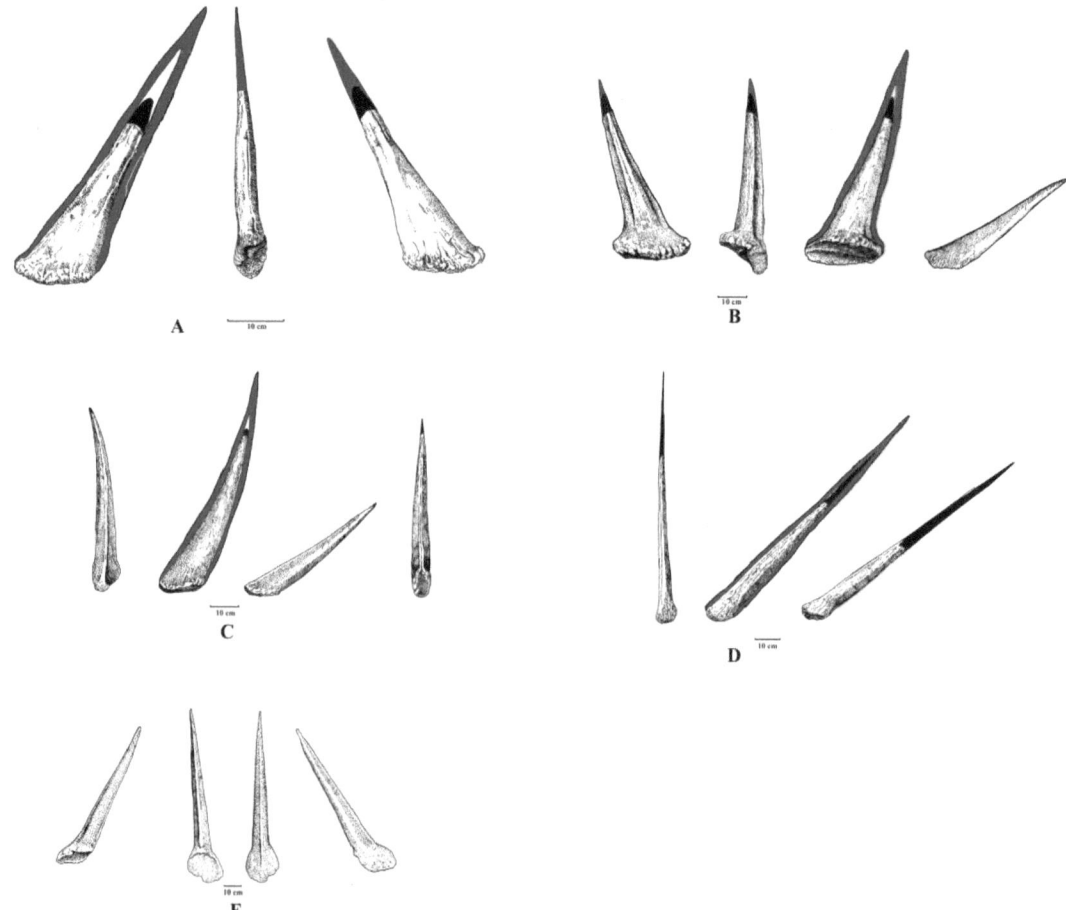

Figure 3). Tail spikes of *Stegosaurus*; A) First caudal spike of *Stegosaurus stenops*; B) Three view of the first caudal spike of *Stegosaurus sulcatus* and dorsal view of the second spike; C) First and second caudal spike of *Stegosaurus ungulatus*; D) Two caudal spikes of *Alcovasaurus* (*Stegosaurus*) *longispinus*; E) The caudal spike I bought at the Tucson Rock/Fossil show in 2005.

Family: STEGOSAURIDAE Marsh, 1877

Genus: *Alcovasaurus* GALTON, & CARPENTER, 2016
= *Natronasaurus* ULANSKY, 2014a (*nomen nudum*)
A. longispinus (GILMORE, 1914) GALTON, & CARPENTER, 1914 (Type)
= *Stegosaurus longispinus* GILMORE, 1914
= *Stegosaurus altispinus* GILMORE 1914 (sic)
= *Natronasaurus longispinus* (GILMORE, 1914) ULANSKY, 2014a (*nomen nudum*)

Subfamily: STEGOSAURINAE Marsh, 1877 (sensu ABEL, 1919)
= Family: HYPSIRHOPHIDAE Cope, 1898

Genus: *Diracodon* MARSH, 1881
D. laticeps MARSH, 1881 (Type)
= *Stegosaurus laticeps* (MARSH, 1881) HENNING, 1915

Genus: *Hypsirophus* COPE, 1878
H. discurus COPE, 1878 (Type)
= *Stegosaurus discurus* (COPE, 1878) HENNING, 1915
= *Hypsirhophus seeleyanus* COPE 1879 (*nomen nudum*)
= *Stegosaurus seeleyanus* (COPE 1879) HENNING, 1915 (*nomen nudum*)

Genus: *Stegosaurus* MARSH, 1877
S. armatus MARSH 1877 (Type)
S. stenops MARSH, 1887
= *Diracodon stenops* (MARSH, 1887) BAKKER, 1986
S. ungulatus MARSH 1879
= *Stegosaurus duplex* MARSH 1887
S? affinis MARSH 1881 (*nomen nudum*)
S. sulcatus MARSH 1887
S. Nova? BAKKER

Bibliography

Galton, P. M., and Carpenter, K., 2016, The plated dinosaur *Stegosaurus longispinus* Gilmore, 1914 (Dinosauria: Ornithischia; Upper Jurassic, western USA), type species of *Alcovasaurus* n. gen.: Neues Jahrbuch für Geologie und Paläontologie, Abhandlungen, v. 279, n. 2, 185-208.

Gilmore, C. W., 1914, Osteology of the armored dinosauria in the U. S. National Museum with special reference to the genus *Stegosaurus*: Bulletin of the United States National Museum, v. 89, p.1-140.

Ostrom, J. H., and McIntosh, J. S., 1966, Marsh's Dinosaur, the Collections from Como Bluff: Yale University Press, 388pp.

Ostrom, J. H., and McIntosh, J. S., 1999, Marsh's Dinosaurs the collections from Como Bluff: Second edition, Yale University Press, 388pp.

Ford, T. L., 2006, How to Draw Dinosaurs. Con- or Pro-boscis Diplodocids or Trunk-aided diplodocids, Science Fact or Fiction. Prehistoric Times, n. 78, p. 20-21.

Chapter 6

Con- or Pro-boscis Diplodocids or Trunk-aided diplodocids, Science Fact or Fiction?

It has been speculated that because of the placement of the naris of diplodocids that they may have had a proboscis like an elephant. Larry Witmer and his crew are working on this topic, but their work isn't finished yet and I will not be using their information. Bakker has toyed with the idea of a proboscis diplodocid in his book, The Dinosaur Heresies. This has always looked funny and wrong to me.

The placement of the naris in diplodocids has confused paleontologist since the very first discovered skull. Because the naris is above and in between the eyes on the top of the skull one idea was they used their nose like a snorkel keeping their body under water and breathing through the top of their head. Once the dinosaur renaissance started, swamp dwelling sauropods was shown to be incorrect; no more swamp dwellers - no need for snorkels. Not to mention the original authors of the submerged sauropod never took into account the weight of the water on the body of the sauropod and whether or not it could actually breath when it was under water, which it couldn't.

Since that theory was dismissed, why do diplodocids have the naris on top of the skull? Another theory was it had a trunk or proboscis like an elephant or tapir. Even though diplodocids have a dorsally placed nasal, it is different in shape and position than those of elephants and tapirs. In the mammals the naris is facing more forward and has a large naris. This allows the proboscis more room for attachment. In diplodocids the nasal is lower and does not face as far forward as in elephants. This doesn't allow much room for attachment of a proboscis.

If there was a proboscis it wouldn't have looked like an elephant. The surface on the 'face' of the skull is wide and the naris are small. If it did have one, it may have been thinner and wider than an elephants. Witmer is working with the soft tissue of modern mammals then extrapolating that to dinosaurs. I'm not sure if I agree using mammals as a template for dinosaurs is a good idea, but then there is no other group of animals that that can be done with today.

But what purpose would a trunk be to a sauropod? Mainly to gather food to its mouth, but would it really be beneficial? They already have a long neck so getting to food (from higher branches or deeper into the tree) so a trunk wouldn't help. The teeth are pencil-like in diplodocids and good for nipping. A strong large tongue would be more beneficial.

In a recent article by Knoll, Galton and Lopez-Antonanzas (Paleoneurological evidence against a proboscis in the sauropod dinosaur *Diplodocus*: Geobios, v. 39, p. 215-221) studied the braincase of sauropods and elephants and came up with the conclusion that they couldn't have had a trunk. Basically, elephants have a large facial nerve that emerges from the brain. A branch of this nerve and also of the trigeminal nerve unite to form the proboscidal nerve which supplies the muscles of the powerful and complex motor system of the trunk. *Diplodocus* on the other hand has a small or nearly non-existent facial nerve. This means they didn't have the nerves for a trunk. Also, there are no muscle scars on the skull to indicate a trunk. One of the reason's why elephants have a trunk is because they have a short neck, and some mammals with short necks have a proboscis which allows them the 'reach' they need to gather food. Since sauropods have a long neck they wouldn't have had a reason for a trunk.

It has been theorized that diplodocids did have a split nasal with a cartilaginous septum. There is a large splint of bone anterior of the nasal and a slight medium division on the posterior portion of the nasal, this could have supported a cartilaginous septum. I see no reason why diplodocids didn't have a 'fleshy' nose to some extent. They'd have to have been able to close their nose when they stuck their head into deep foliage or they'd have gotten food stuck in their nose. Could diplodocids been able to use the placement of the naris to breath and swallow at the same time to create a 'suction' in order to eat more foliage?

Editors Note: I corrected/expanded on sauropod noses via Witmer's research in a latter article, PT 96, 2011, and Chapter 24 of this book.

Bibliography

Bakker, Robert T. 1986. The Dinosaur Heresies, New Theories Unlocking the Mystery of the Dinosaurs and their extinction. William Morrow and Company, Inc. New York: 481pp.

Knoll, F., Galton, P. M., and Lopz-Antonanzas, R., 2006, Paleoneurological evidence against a proboscis in the sauropod dinosaur *Diplodocus*: Geobios, v. 39, p. 215-221.

Ostrom, J. H., and McIntosh, J. S., 1966, Marsh's Dinosaur, the Collections from Como Bluff: Yale University Press, 388pp.

Ostrom, J. H., and McIntosh, J. S., 1999, Marsh's Dinosaurs the collections from Como Bluff: Second edition, Yale University Press, 388pp.

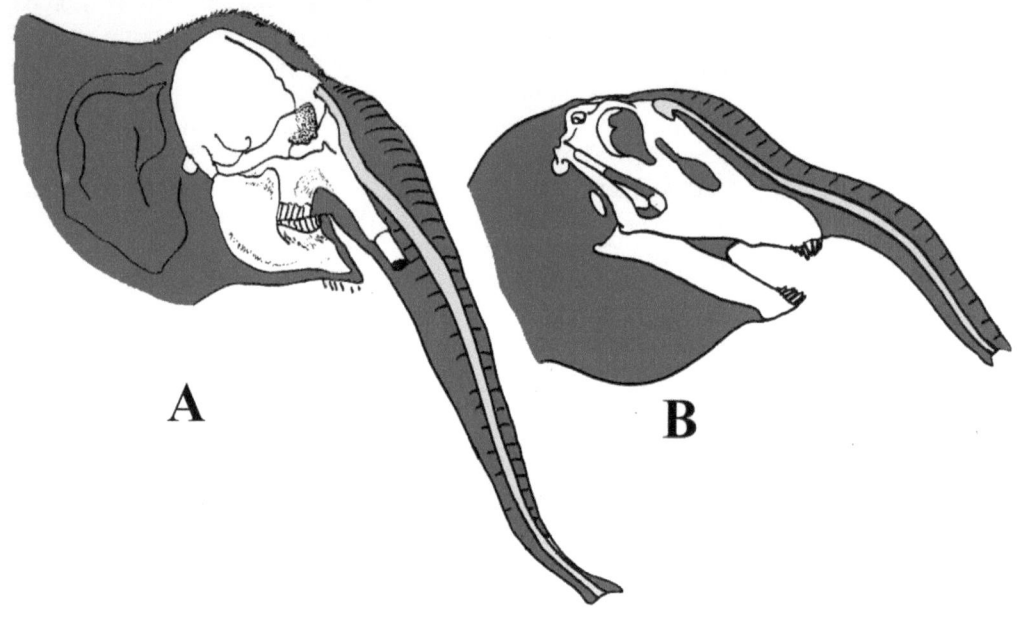

Figure 1). Skull of Asian Elephant A); and of a hypothetical *Diplodocus* B). This illustration shows the proboscis and the nasal conduit

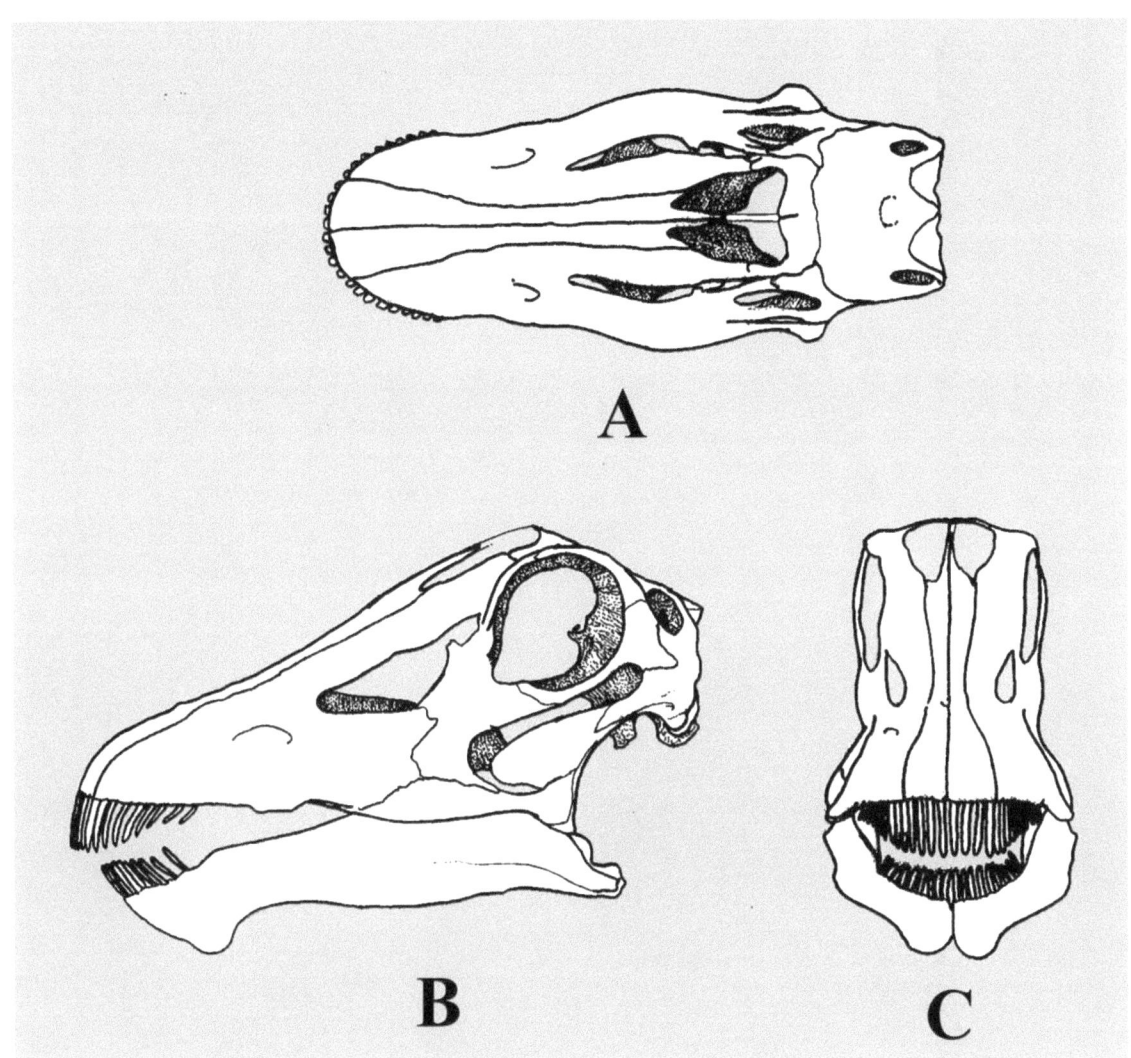

Figure 2). Skull of *Diplodocus* in dorsal A); lateral B); and anterior C) views.

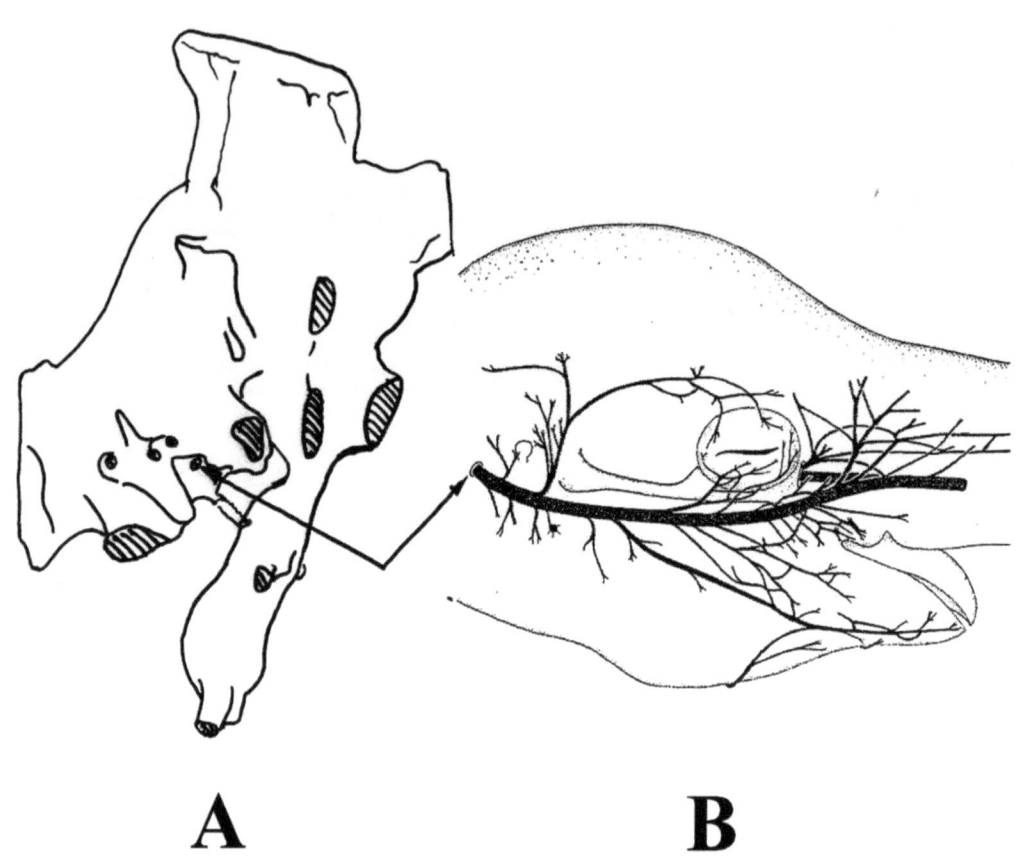

Figure 3). Braincase (endocast) of *Diplodocus* with arrow pointing to the diminutive size of the facial nerve A); and a fetal *Loxodonta* (African elephant) showing how large the facial nerve is in a fetus.

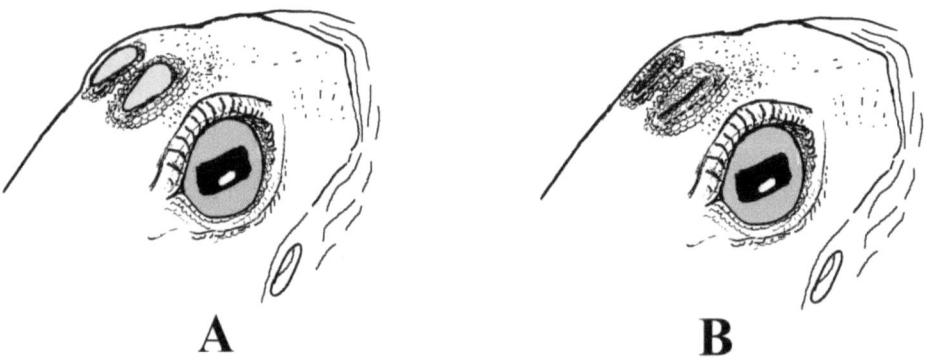

Figure 4). Illustration showing an open naris A) and a closed one B) of *Diplodocus*.

Ford, T. L., 2006, How to Draw Dinosaurs er Ammonites: Ammonites are not Nautiloids, part 1. Prehistoric Times, n. 79, p. 20-21.

Chapter 7

Ammonites are not Nautiloids, part 1

This time around I'm not talking about dinosaurs but a group of invertebrates that I find fascinating, Ammonites. My big beef that I have with artist depicting ammonites is that they make them look like a Nautilus.

Ammonites are NOT nautiloids. They are distant cousins, twice removed. They don't have the same body plan or morphology other than both having a shell, and even that is different enough to identify each by their shells. This will be an anatomy lesson to distinguish the two groups. I was debating whether or not to make this a two parter or not to make is a two parter and I've decided to go ahead and make this article a two parter. The first part will be on the anatomy the second part will be on the different shell shapes. Also, I would like to thank Neal Larson and my friend Gerry Alverado who looked over this article for me.

My interest on the subject started when I bought Neal Larson's book, *Ammonites and the other Cephalopods of the Pierre Seaway* way back in 1997. I was so enamored by the different ammonites that I just had to draw one. I drew *Didymoceras stevensoni* (on page 55 if you have the book). The next time I saw Neal I showed him my drawing. He liked the shell but pointed out I needed to correct a few things. For one thing, it needed to have spines. It turns out the spines just about never fossilize (or are broken off). The other correction he pointed out was that I drew it like a nautilus and it should be like an octopus. It is because of that conversation that I write this article.

The spines were easy, but I need to change the body. This was before I had photoshop so I redrew the body, cut and pasted (manually) onto a copy of the shell. In fact, I drew 4 different bodies to simulate the body being retracted into the body chamber (Figure 1). He liked that one along with my drawing of *Didymoceras binodosum* (one of my favorite ammonite drawings) (Figure 5).

Nautiloids are in the cephalopods group (class/ clade, what ever) Tetrabranchiata (four-gilled). Ammonites are most likely in the group Dibranchiata (two-gilled) which contains the squids, octopi, belemnites and cuttlefish and NOT in the group with *Nautilus*. This is based on all of the little things that we find with ammonites and their similarity to the dibranchiates. They have been mistakenly placed in the tetrabranchiates along with *Nautilus*, because people have always assumed that they are like *Nautilus*. Therefore, Ammonites should be depicted like an octopus or squid in a shell.

The first cephalopods appear in the Late Cambrian. By the Ordovician, nautiloids (both straight and coiled) were quite abundant. Ammonites first appear in the Late Silurian, having split away from some of the straight "Nautiloids" (which aren't really true nautiloids, many of them are squid like) in the Bactridiae family. For some reason ammonited died out the same time as the dinosaurs where the nautiloids lived on.

Both have a shell made up of aragonite and in both the shell is manufactured by mantle tissue. In the shelled cephalopods (ammonites and nautiloids) the mantle surrounds the visceral sac (within the body chamber) and possesses strong muscles required for respiration and contraction for propulsion.

The shell is made up of two parts; the large body chamber which obviously housed the soft tissue and the chambered section called the phragmocone. I have seen some cut ammonites that have small shells of different invertebrates in where the body chamber would have been. The individual chambers of the phragmocone are called camerae. The beginning of the phragmocone contains the ammonitella or hatchling shell. The animal grew one chamber at a time. Each chamber is separated by a wall or septum and is joined to other chambers by a tube called the siphuncle. The siphuncle in nautiloids is in the center of the shell and in ammonites the septum has several siphuncles along the walls. The siphuncle removed any fluids (by osmosis) that might accumulate in the chambers as the animal grew (Figure 2). The phragmocone acted in the same way as a swim bladder in fish, thus providing neutral buoyancy. They maintained neutral buoyancy throughout its life. Despite popular belief, shelled cephalopods did not act like a submarine. They did not move gases or liquid from chamber to chamber in order to go up and down in the water column. Because they had neutral buoyancy they moved the siphon and squirted water in the direction they wanted to go (Figure 2).

34

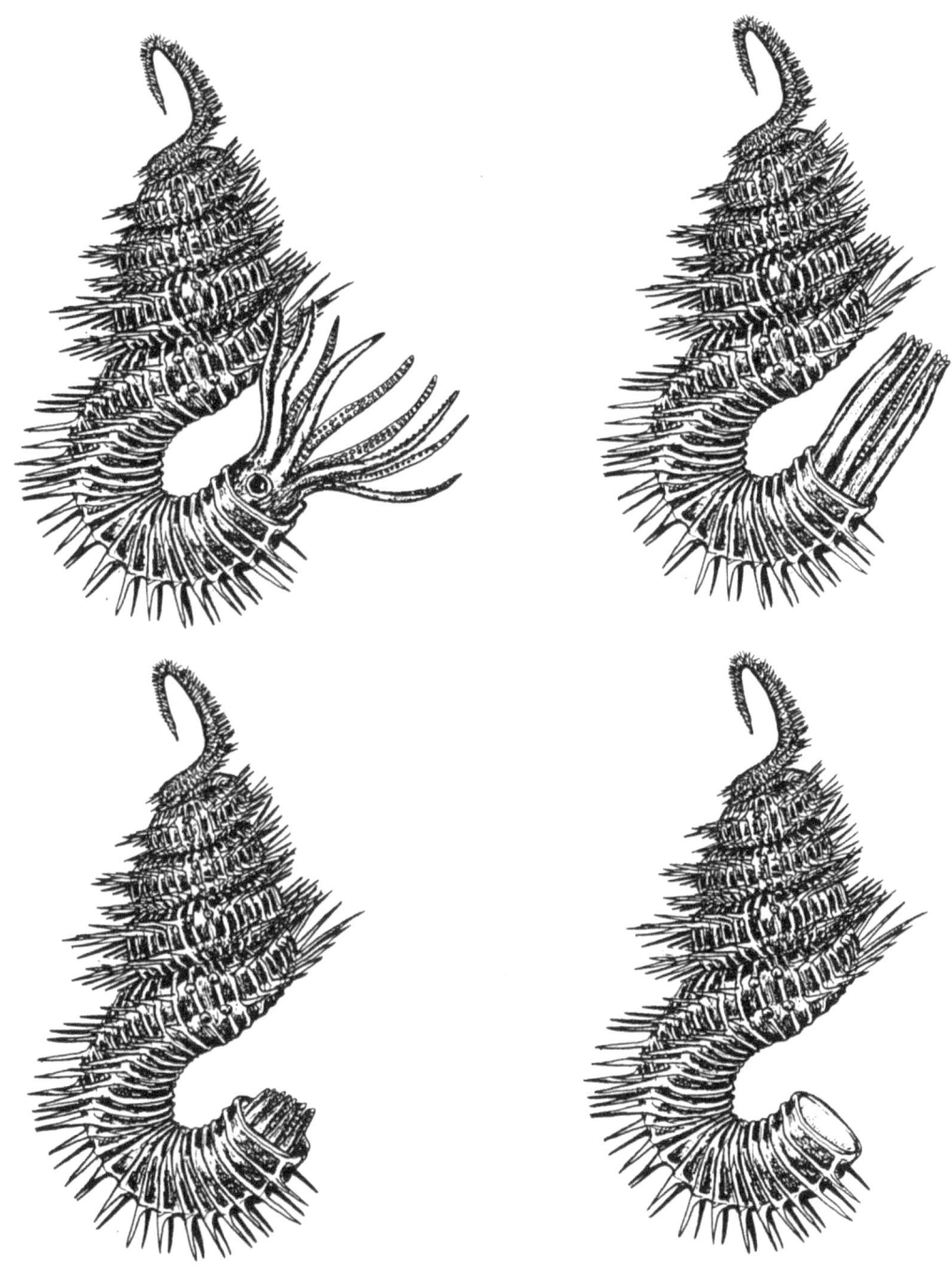

Figure 1); 4 images of *Didymoceras stevensoni* withdrawing into its body chamber.

    The outer shell is made up of the outer prismatic layer; the outside dull layer, and the inner nacreous layer, 'mother of pearl' layer (which when fossilized can be misinterpreted as opal. Ammolite is the fossilized mother of pearl that is sold in Gem and Mineral shows). Lying directly under the shell is the septal layer which contains those many 'squiggly lines' or better known as the suture pattern. The suture pattern is different for every family, genus and species of ammonites and is used to help in identification and description of each species. Nautiloids lack the suture layer. In between each layer are conchiolin layers. All these layers are deposited by the mantle in the same way that modern mollusks create their nacreous shell. (Figure 2)

Unfortunately, no soft body parts of ammonites have ever been found.

I won't be going into the internal organs but concentrate on the parts that are important to the artist; the outer shell, eyes, arms, etc.

In nautiloids the outer shell has a large black area (just above the body chamber) where the outer hood lies. It conforms closely to the size and outline of the shell aperture. The larger the area, the larger the hood. The hood is lacking in the octopi, squid and cuttlefish and was also more than not, lacking in the ammonites.

Nautiloids have 90 or more tentacles. There are 3 different tentacles, the ocular, digital and labial. There are 2 pairs of ocular tentacles, they are small and are near the base of each eyestalk, one in front, and one in back.

The labial tentacles form an inner ring, and the digital are the outer ring.

Some of the tentacles are olfactory, others respond to touch, yet most are prehensile. Each tentacle has a proximal sheath and a distal long noodle-like cirrus. (Figure 4) The cirrus can be completely withdrawn into the sheath in the ocular and digital cirri. The cirri do not have suckers.

But the cirri do act like suckers, both adhesive and prehensile. It carries many annular grooves and ridges. The latter are more prominent on the inward face and this projecting part of each ridge serves as an adhesive organ and acts in a similar way as suckers. Also, they produce a sticky mucus to help hold onto things.

The eye of the nautilus is different from those of octopi and squid. It is extremely primitive in structure. The eye is on a stalk and has a pin point hole on open water-filled pits. They have a retina but lack an iris, cornea, lens or eyelid. It is square and is split from the center of the eye to the lower edge. In octopi, squid and cuttlefish the eye is the most complex in any invertebrate and is very similar to humans. They have image forming complex eyes. How they focus is by moving the lens back and forth as appose to the iris changing shape. The squid has the largest eye to body size of any animal. The eye is large and the 'iris' in squids are oval and rectangular in octopi and cuttlefish.

No shelled cephalopod has an ink sac.

Cephalopods move by squirting water through the hyponome or siphon (which also has the gills at the inside edge). In the nautilus the siphon is composed of two converging folds while the other cephalopods are fused into a tube. Incidentally the hyponome is remnants of the 'foot' from their early ancestors as are the tentacles/arms. They can manipulate the hyponome in any direction. (Figure 3)

The cephalopod beak is made up of two parts and in some it looks like parrots.

Ammonites it is believed to have 4 to 5 pairs of arms like octopi and squids. If they are like squids then they would have had 2 long tentacles.

They would also have had suckers and hooks on the arms. In octopi and squids there are 2 rows of suckers on the arms and in the tip of the squid there are many more suckers. Each sucker can act independently or in unison with one another. In the octopi they are round, in squids the suckers are on stalks and have a ring of chitinous, centrally directed teeth. In some species it may be reduces to one large hook (Beleminites fossils have been found with the large hooks and impressions of arms. Several marine animals have been fossilized with beleminite hooks in the stomach region). They can also grow a new arm if it is diseased or torn off. The arms radiate around a 'beak'.

Ammonites lack a hood and should not be illustrated with one. Some octopi have 'loose' arms, and others the arms are connected by a partial web or fully webbed.

The shell shape differs more in ammonites than nautiloids. Both have a straight shelled forms and coiled forms, but in ammonites some of these coiled forms came in bizarre shapes. Some tightly coiled, others loosely, some have ridges, others have horns on the ridges. In some ammonites the shell nearly splits the body in half. Both could retract the body into the body chamber. In nautiloids the hood would act like a wall, but ammonites could completely withdraw the body into the body chamber.

Cephalopods communicate by changing their color or color patterns; to ward off animals or attract mates. This is done by chromatophores in the 'skin'.

Octopi and cuttlefish can change their skin texture from smooth to rough or bumpy. Ammonites may have been able to do the same thing.

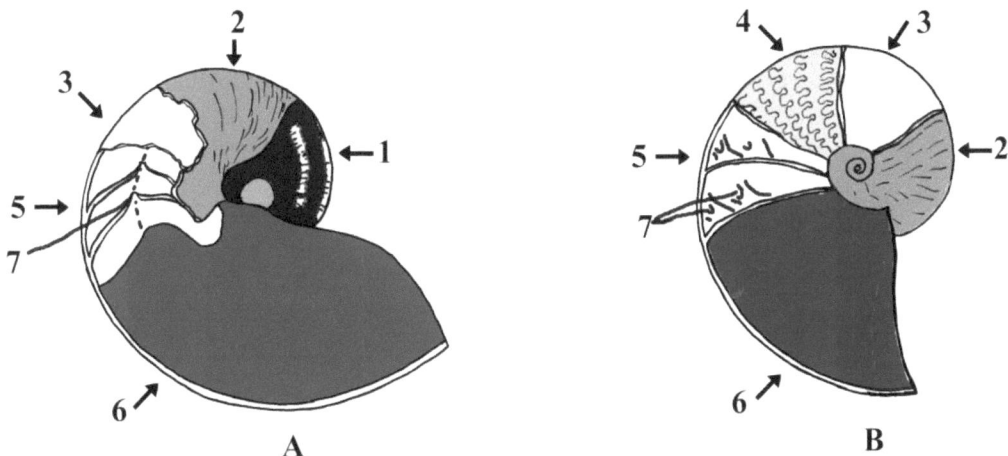

Figure 2); Shells of a Nautilis A); and an Ammonoid B). 1 black film; 2) prismatic layer; 3) nacreous layer; 4) septal layer; 5) camerae surrounded by the septum; 6) body chamber; 7) siphuncle.

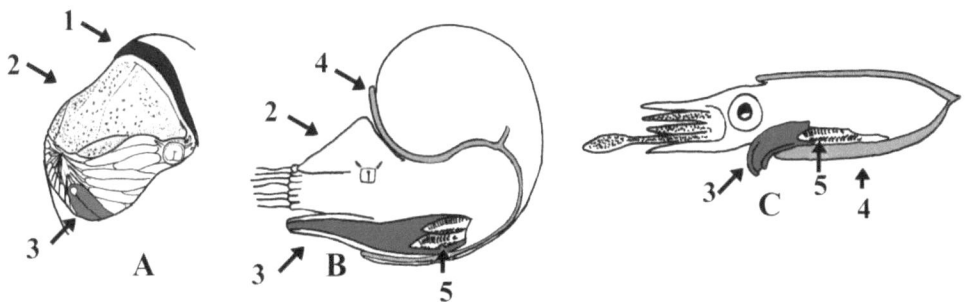

Figure 3); Cutaway diagram of a nautiloid B and a squid C and a front view of a nautiloid A; 1) black film; 2) hood; 3) siphon; 4) mantle; 5) gills.

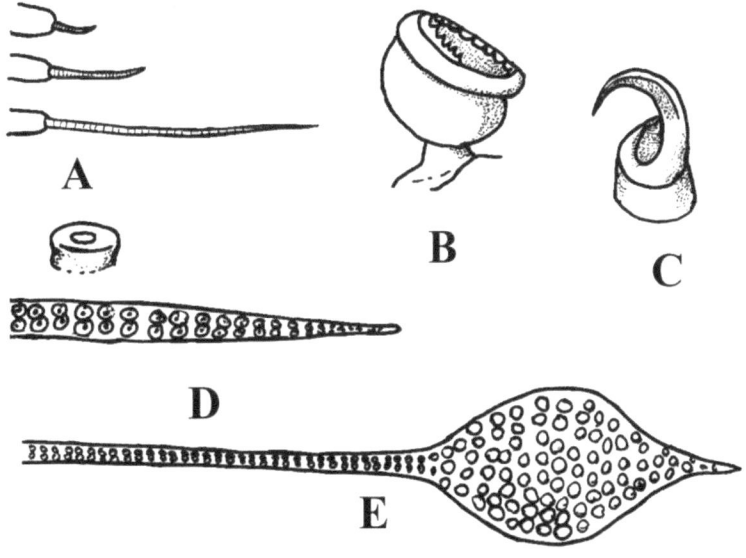

Figure 4); Arms and tentacles; A) arms of a *Nautilus* in 3 stages of protrusion of the cirrus (arm) and sheath; B) sucker of the squid *Todarodes sagittatus* with horny ring of teeth and stalk; C) tentacular sucker of *Onychoteuthis banski*; D) generic sucker of an octopus and arm; E) generic tentacle of a squid.

In conclusion, ammonites are not nautiloids and should not be depicted as one. They should have 4 to 5 pairs of arms with two rows of suckers. The suckers can be like those of an octopus or squid. You can either put a squid like eye or octopi like eye in your ammonite, not a *Nautilus* eye.

Figure 5); *Didymoceras bindosum* with webbed arms.

Ford, T. L., 2006, How to Draw Dinosaurs er Ammonites, Ah make that cephalopods, Ammonites are not Nautiloids, part 2 (and other fossil cephalopods). Prehistoric Times, n. 80, p. 18-19.

Chapter 8

Ammonites are not Nautiloids, part 2 (and other fossil cephalopods)

There are 2 living groups of cephalopods, the Nautiloidea, and the the Coleoidea. The coleoidea consist of the Octopus, Squid and Cuttlefish. The coleoidea did not replace or out compete the ammonites. For nearly all the Mesozoic they lived side by side. Nearly all the coleoidea have some sort of hard 'shell' inside their body with only the octopi not having one. They also have different names for this structure; cuttlebone, gladius, pen etc. (though some names do overlap). These structures are all made up with aragonite (which is a calcium carbonate). No cephalopods live in freshwater, living or extinct.

The bodies are divided in different sections. The squid and octopus has the arms, head, and mantle, the cuttlefish has mantle and arms. Nearly all also have ink sacs. (Figure 1).

Coleoidea
Cuttlefish (living, 119 species)

Cuttlefish belong to the Sepiida. They have an inner shell (cuttlebone or sepion) (Figure 2). The cuttlebone is made up of calcium carbonate. It is very porous and has chambers and is what the cuttlefish uses for bouncy. It does this by changing the gas-to-liquid ration in the chambers of the cuttlebone. The cuttlebone is made up of an outer cone, and an inner one with a very small rostrum or spine. Each cuttlefish has a distinctive cuttlebone. We use cuttlebones for parakeets to chew on and is used by jewelers and silver smiths for casting small objects. But don't worry, cuttlefish only live about a one to two year and are very abundant. They have large eyes, 8 arms and like squids have 2 tentacles. The fin covers the complete length of the mantle. One way they swim is by undulating the fin, they also have a siphon and are able to squirt water to propel them. Cuttlefish are the chameleon of the seas and can blend into their surroundings at astonishing speeds. They are very social and if you ever get a chance to see them in an aquarium I highly recommend it. Cuttlefish eat fish, mollusks, crabs, shrimps and other cuttlefish. They spend some of their time buried in the ocean floor. No cuttlefish are known from the oceans around the United States.

The earliest record of cuttlefish is from the Lower Jurassic. Several have been found in the Solnhofen and the Middle Cretaceous Lebanon. In the Solnhofen the Cuttlefish grew to more than a meter in length and are the largest invertebrate at that time. Some have large bodies and short arms, others had arms as long as their body. Some bodies are wide others are thin.

Sepiolida (bob tail squid) (living, 70 species)

These small cephalopods are related to cuttlefish. They don't have a cuttlebone and there are no known fossils (Figure 2).

Rams Horn Squids (Spirula) (living, 1 species)

This group of cephalopods are a unique group. They have a curled inner shell in some ways looking like that of an Ammonite. Like a squid they have 8 arms and 2 tentacles. The shell is loosely curved and is not tightly packed as in ammonites. The shell helps them maintain a life position with the head and arms pointing downward (Figure 2). There is a Late Jurassic form *Gramadella* from France.

Belemnites (extinct)

Belemnites had a body shape similar to squids, but they are not related (Figure 3). In Germany they are also known as Thunderbolts. They also had a hard shell that was inside their body. The 'shells' are very distinct, and some Belemnites are used as biozones, meaning that only certain types are known from a specific time period and when that belemnite is found the age is known. I won't be getting into shell shape because they wouldn't have been seen, though I will explain a bit of it. The 'shell' is made up of the phragmocone, which is chambered and has a sphencule, and a large solid conical rostrum or guard. The rostrum is the most commonly found part of the 'shell'. Belemintes range from less than 10 centimeters (3.9 inches) to over 3.9 meter (13 feet), the largest being *Megateuthis* (from the late Jurassic of Germany) with a rostrum of over 70 centimeters (27 inches).

The first belemnites are known from the Carboniferous (Mississippian) and died out at the same time as the ammonites. Belemnites had 'hooklets' on their arms. These hooklets have been found in the gut contents of *Tanystropheus* (which is one of the reasons I believe it was a marine animal and not a terrestrial one), Ichthyosaurs, Plesiosaurs and some fish. There was unique trace fossil that for decades mystified paleontologist (*Plesioteuthis prisca*). It was a rosette of 8 appendages. It was later found out that it was from belemnites 'standing' on their arms.

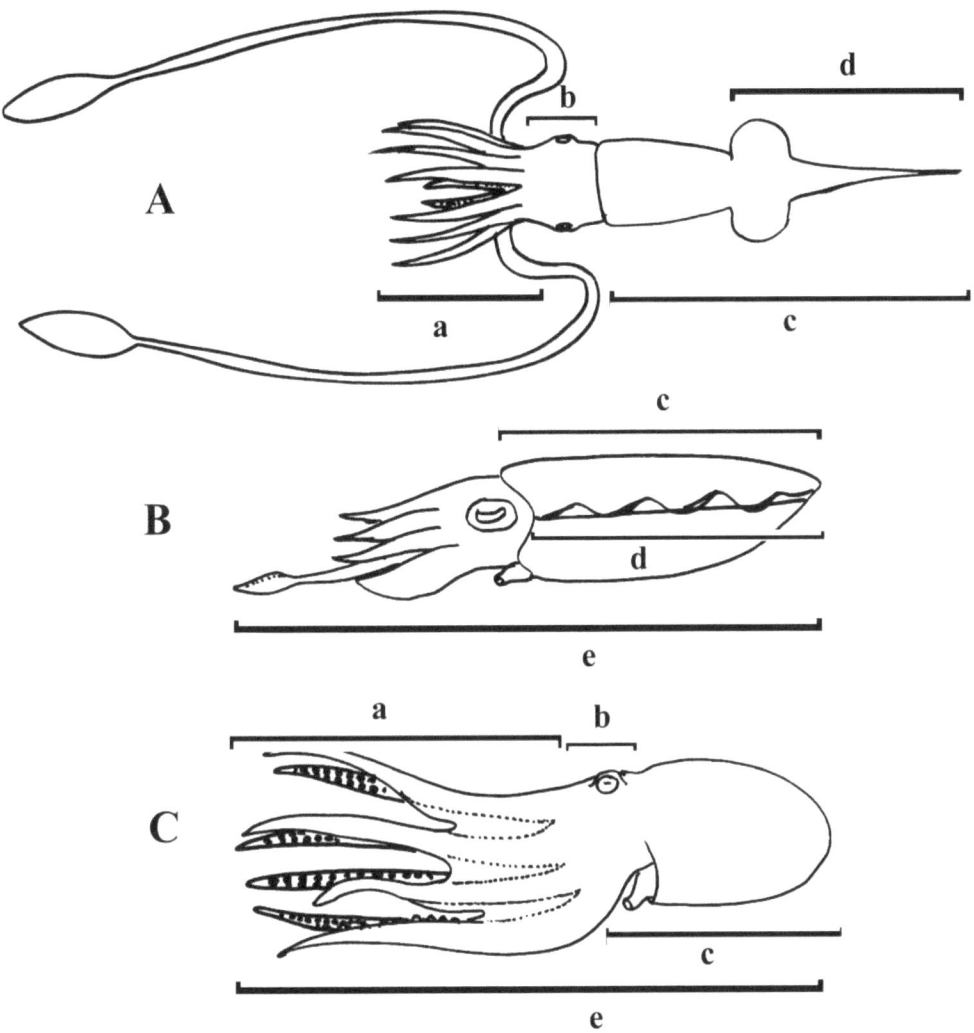

Figure 1); Schematic illustration of cephalopods; A) Squid; B) Cuttlefish; C) Octopus. a) arm length; b) head; c) mantle length; d) fin length; e) total length.

Squids (Teuthida) (living, 298 species)

Squids have been around since the early Devonian and were more abundant during the Jurassic to Cretaceous, but not nearly as abundant as they are today. Squids have a hard chitinous supporting structure called the gladius (or pen) (Figure 2). This structure has a long rhachis and a spoon like structure. The gladius protects the inner organs. The gladius is the main part of the squid that does fossilize. Large gladius of some Niobrara squids (*Niobarateuthis* and *Tusoteuthis*) indicate some of the squids during that time grew as large as *Archeoteuthus* (the giant squid). But since the Niobrara seas were shallow these large squids lived a different life than modern Giant Squids. Some gladius have tooth marks from mosasaurs and sharks.

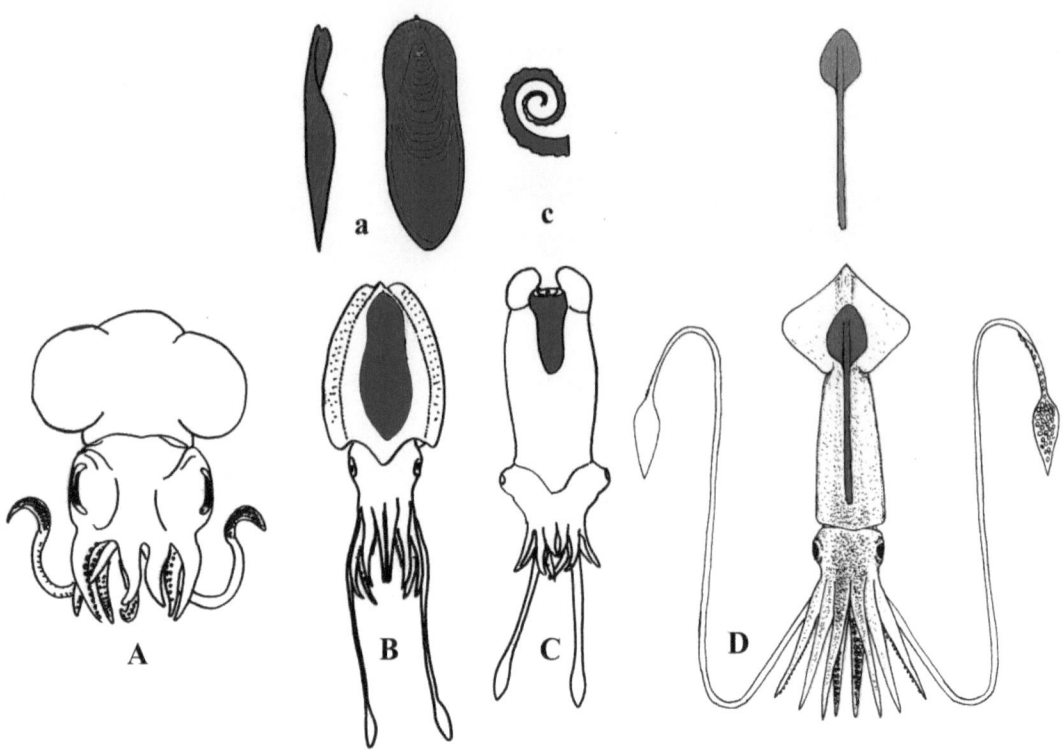

Figure 2). Cuttlefish and squids; A) Sepiolidae (bobtail squid); B) Cuttlefish; C) Spirulidae (rams horn squid); D) Squid. Position and type of inner shell; a) cuttlebone/ sepion; b) shell; c) gladus/pen.

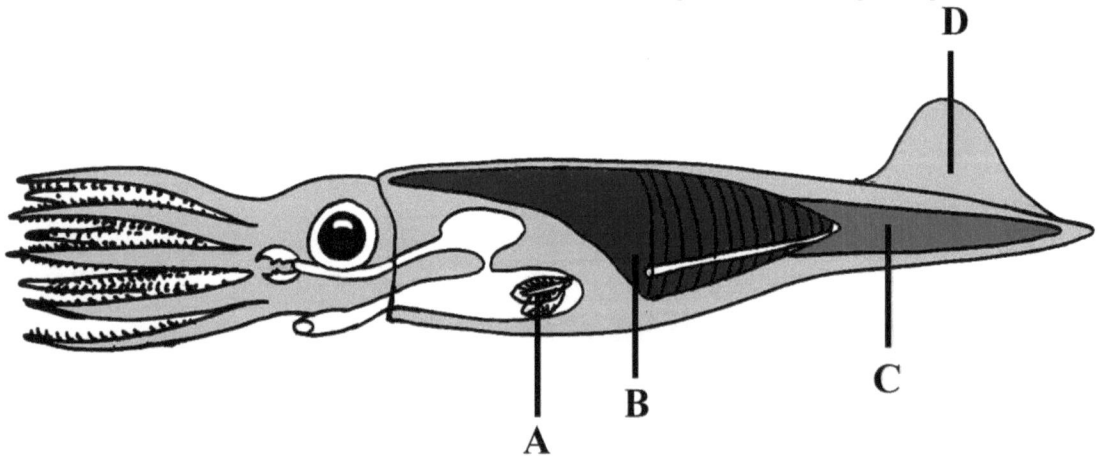

Figure 3) Diagram of a belemente; A) Gills; B) Phragmocone; C) Rostrum; D) Fin.

Octopoda (living, 298 species)
Vampyromorpha (living, 1 species)

The classification of octopoda is as confusing as that of birds. The Division, grade or whatever it's called has gone from Neocoleoidea, Vampyropoda, and back with the superorder changing names several times, Vampyromorphoidea, Octobrachia, Octopodiformes and Octobrachimorpha. The basic gist is there is the 'Vampyomorpha and Octopoda. Vampyomorpha have a chintious gladus that is thin (Figure 4), and forms a broad plate, in Octopoda the shell is vestigial or has small cartilaginous rods.

Fossil octopi are few and far between (Figure 5). They seldom fossilized and those that do are just impressions or stains. The earliest octopus is from the Mazon Creek of Illinois. *Pohlsepia mazonensis* was found in the brackish water beds and was a very small animal with 10 arms. It is very small, only 25

42

millimeters by 35 millimeters and has two fins on its mantle like those of modern Cirrate octopuses. The next octopus comes from the middle Jurassic (lower Callovian) of France (*Proteroctopus ribeti*). It has 8 long arms and traces of suckers and also has two fins and was about 4 inches long. The next fossil octopus comes from the Middle Cretaceous deposits of Lebanon and is known from several specimens, *Palaeoctopus newboldi*. It has 8 long arms, and also paired fins and was also a small octopus. It is argued that some if not all the fossil 'Octopoda' are actually Vampyomorpha which would mean there are no 'real' octopoda know from the Mesozoic.

Vampyromorpha are known in the fossil record, a gladis from the Late Carboniferous and several *Muensterella scuttellaris* are known from the Solnhofen of Eichstatt Germany. *Vampyronassa* is known from the Middle Jurassic of France (Figure 4).

A few Tertiary octopi have been suggested from drill holes in some scallops.

The paper nautilus isn't a nautilus (Argonauts), but belongs to the octopus family (Figure 5). Females make a 'shell' for their egg casing. These shells are not like ammonite or nautilus shells and do not have chambers. The females have specially adapted arms that form the shells. They do not live in shells like the nautilus. There are also several Argonaut shells known from the Miocene.

So when making a Mesozoic ocean scene feel free to also include cuttlefish, squids, octopi or belementies.

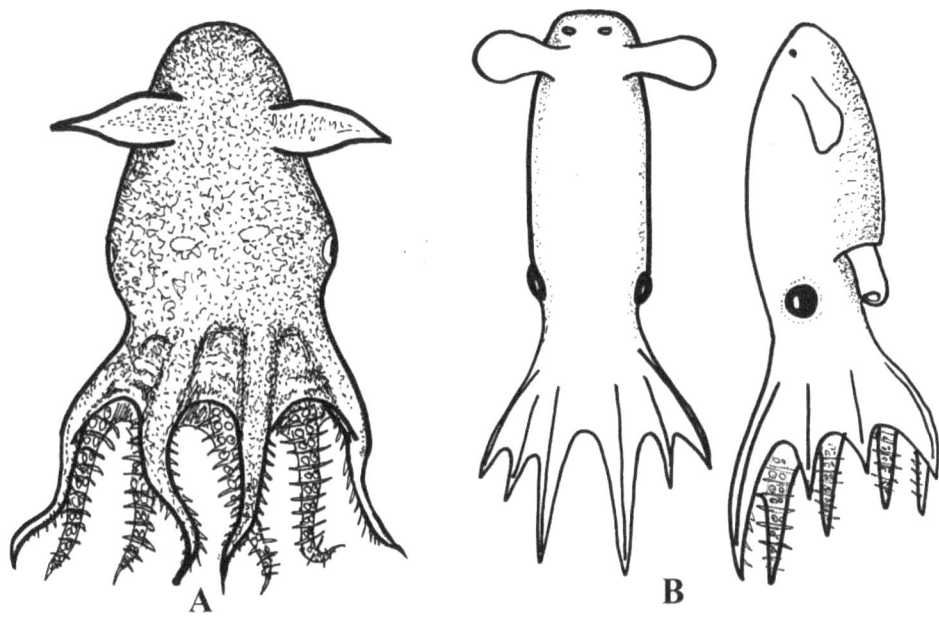

Figure 4) Vampyromorpha; A) Living *Vampyroteuthis*; B) Fossil *Vampyronassa rhodanica*.

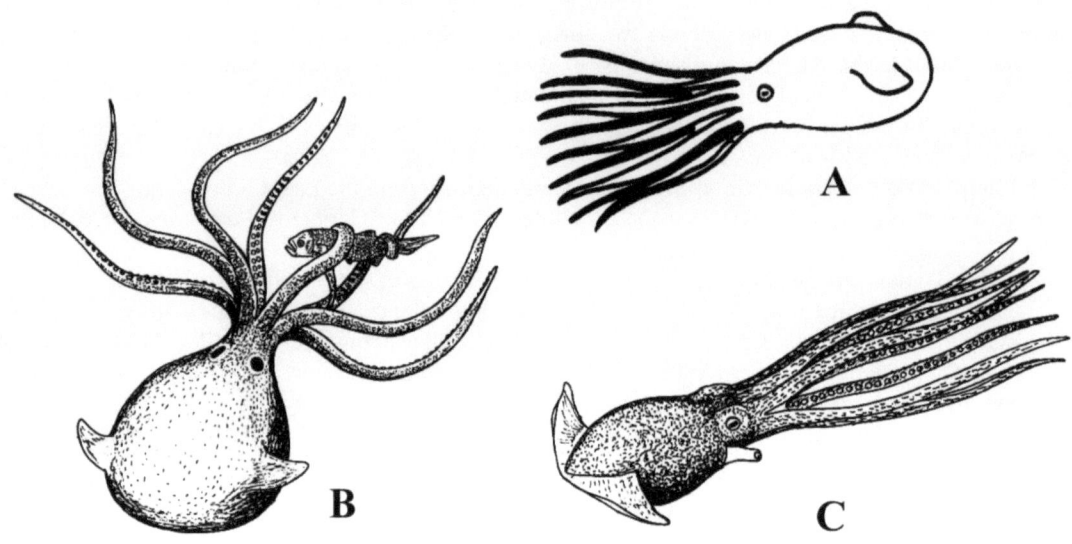

Figure 5) Fossil Octopi; A) *Pohlsepia mazonensis*; B) *Palaeoctopus newboldi*; C) *Proteroctopus ribeti*.

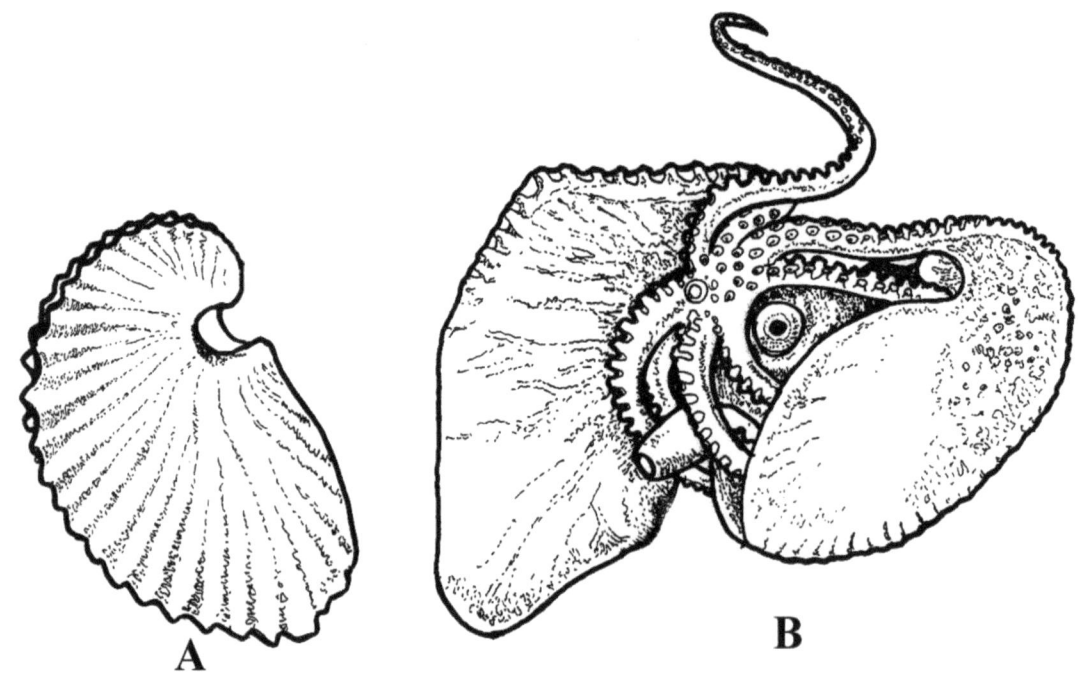

Figure 6) Paper nautilus; A) Egg casing; B) female.

44

Ford, T. L., 2007, How to Draw Dinosaurs, A little cartilage goes a long way. Prehistoric Times, n. 81, p. 18-19.

Chapter 9

A little cartilage goes a long way (Back to dinosaurs...)

A new paper has recently been published that may have a huge impact on how dinosaurs are depicted not only on paper but also mounted skeletons. All animals have cartilage and it has been speculated on how much cartilage dinosaurs had. Some believe they had a little, others believe a lot. But without skeletal evidence, it is all speculation. Schwarz, Wings and Meyer (*Super sizing the giants: first cartilage preservation at a sauropod dinosaur limb joint:* Journal of the Geological Society, London, v. 164, p. 61-65.) describe such evidence. They reexamined the right humerus of *Cetiosauriscus stewarti*. The distal end of the humerus has a different texture and color from the rest of the humerus. It turns out the distal end has fossilized cartilage. They drilled small holes in the distal end of the humerus to get samples for histological studies and discovered that they went through both cartilage and bone.

The cartilage (light grey) is easily distinguished from the brown bone color (Figure 1). The contact between the bone and cartilage is sharp and the cartilage is poorly bound to the bone. It covers the distal extremity except along the lateral third. On the olecranon fossa the cartilage layer is 3 times as thick as the rest of the surface. The cartilage extends for 11.5 centimeters proximally at the lateral ridge and lateral margin, and 8.5 centimeters proximally in the depressions laterally and medially to the lateral ridge.

The surface of the cartilaginous area is crossed proximodistally with furrows and ridges of 1-2 millimeters with and 2-5 millimeters apart and is best developed in the olecranon depression. The cartilage has circular perforation pores with diameters of 0.5-1.5 millimeters. Interestingly the cartilage also has groves where the muscles and tendons would pass by (Figure 2). They believe with the increased cartilage it would lessen the range of movement of the legs, though I'm at a loss as to why.

Without getting more technical the long and short of it is it is cartilage and they checked for all the distinguishing characteristics to make sure it was cartilage. Oddly none of the other bones associated with it has cartilage. They believe this may be from either a different taphonomical process or pathologic. The bone may have been diseased and produced more cartilage, but it shows no other pathologic features. They compare it with baby hadrosaur bones which have cartilage and they are somewhat similar. But that might just be because of ontogeny.

If what they claim is true, the limb bones had large areas of cartilage on the distal end of the humerus and proximal ends of the ulna and radius (Figure 3). The pelvic and pectoral girdles must have had a lot of cartilage. Just look at the opening in the pelvis, it wasn't just a hole. If this is what all dinosaurs had, it would increase the height of the animal and mean that all the skeletal reconstructions are too short.

At the last SVP (2006) I was standing by Senter and listing to him talk to someone about cartilage and how much theropods had. He said he never compensated it in his research when he looked at the arm movement in theropods and they would have had a larger range of movement at the shoulder. This in itself would dismiss much of what he has published on and how dromaeosaurs and *Archaeopteryx* couldn't raise their arms above the shoulder. This in turn would give them a larger wing stroke.

If true cartilage would have to be take in to account than all dinosaur locomotion studies need to be redone. All dinosaurs would stand taller and theropods would have had a great degree of arm movement and those that flew would have had a large wing stroke.

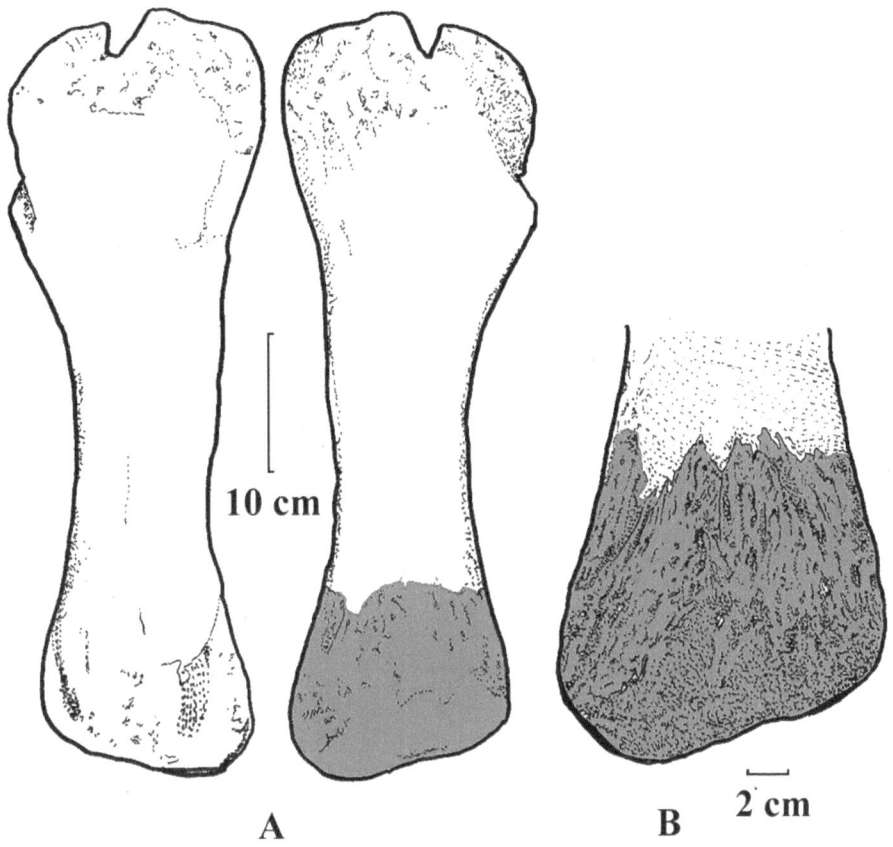

Figure 1): Humerus of *Cetiosauriscus stewarti*; A) in caudal (left) and cranial (right view); B) distal end of the humerus with the cartilage in grey.

Figure 2): Reconstruction of the cartilaginous cap around the humerus, radius and ulna with possible important forelimb muscles.

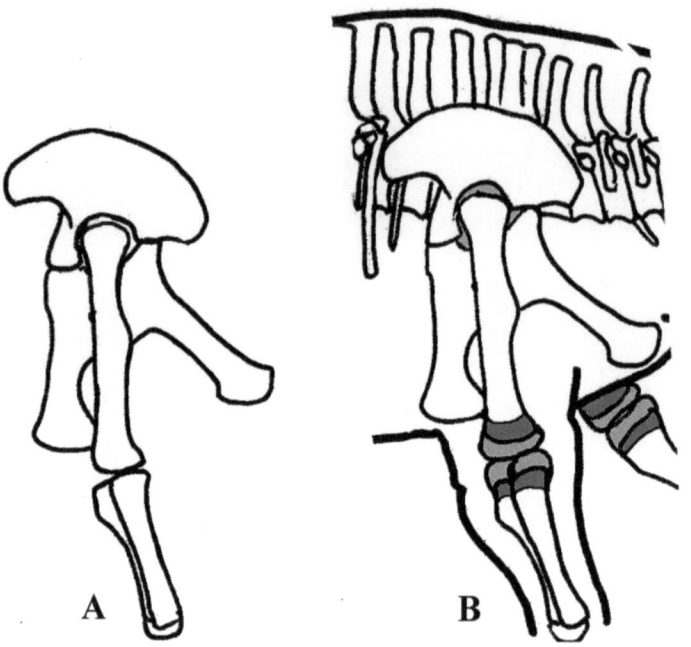

Figure 3): Pelvis of *Brontosaurus* (Yes, I believe it is a valid genus); A) shown typically without the cartilage; B) shown with cartilage.

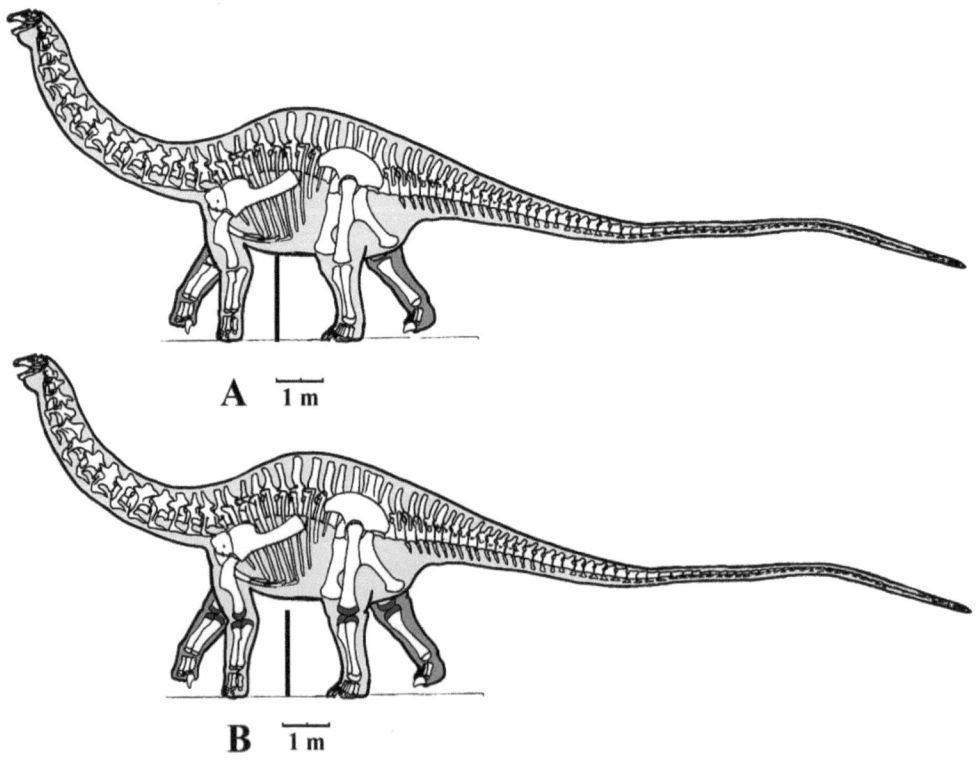

Skeleton of *Brontosaurus*; A) shown typically without the cartilage; B) with cartilage. I have put a solid line indicating the how high the body is from the ground. You'll note that it is higher even though it doesn't look that way off hand.

Ford, T. L., 2007, How to Draw Dinosaurs, Dinosaurs, can you dig them? Fossorial fossils. Prehistoric Times, n. 82, p. 18-19.

Chapter 10

First off I'd like to give my condolences to Riff's family and Mike and his. It was a shock to hear of his untimely death. He will surely be missed.

Dinosaurs, can you dig them? Fossorial fossils.

Actually, I should have said, can they dig? I find it funny, that when a 'modern' ecological niche is found in the fossil record it's amazing or unheard of, but for me I not surprised and am glad it has been discovered. All the modern niches of mammals are the same niches that dinosaurs had, so it shouldn't be a surprise when dinosaur fossils are discovered that filled one of those niches. Some of the niches that dinosaurs aren't supposed to have is aquatic, arboreal, and fossorial. Burrowing dinosaurs has been speculated for a while now though not well received. Bakker (1990) believes *Drinker* was a digger, and Fastovsky et al, (1998) propose *Protoceratops* was a digger and lived in burrows. But recently a new dinosaur has been discovered in a burrow/den. Animals that burrow do so for different reasons; to find food, environmental reasons, escape predators, has dens to live and sleep in, or as primarily dens for rearing young.

Burrows have been found for a variety of fossil invertebrates and vertebrates; crayfish burrows have been found in several locations and eras, lungfish, lepospondyl amphibians, amphibians, small dicynodonts and therapsids, and now dinosaurs.

Oryctodromeus cubicularis has been found in a den. One adult and two juveniles were found which helps to confirm that some dinosaur did have parental care for their young. *Oryctodromeus cubicularis* means "digging runner of the lair". It's a small ornithopod (or Euornithopod, you gotta love those cladistics) hypsilophodontid with an adult size about 2.1 meters, half of which is tail and the body about the size of a coyote. They were found in the expanded end of a chamber (den). The burrow was found in and dried old river floodplain. The burrow is a sloping sinuous tunnel ending with a terminal chamber. About 2 meters long by 30 centimeters wide by 40 centimeters tall. The lithology of the structure is different from the surrounding rock so there is no question on what the structure is. The den is similar to modern animals; rodents, striped hyena (*Hyaena hyaena*), aardwolf (*Proteles crisatus*), gopher tortoise and puffin (*Fratrcula arctica*). The taphonomy, burrow dimensions and anatomic evidence indicate that *Oryctodromeus* was the burrow maker. The possibility of the dinosaurs being placed in the den by a flood is not supported by the lithology. The burrow was a tight fit and the dinosaur was just able to get inside, about the same width of the animal. This is not unusual for a fossorial animal to have the tunnel just big enough for the animal to fit in. *Oryctodromeus* must have kept its femur level and leg bent and moved just by its feet. It is believed they dug with their hands, feet and snout.

The skeletal features that support this dinosaur as being fossorial is its large robust scapula/coracoid, an expanded sacrum, greater attachment of pelvis and sacrum, absence of a pelvic symphysis, and expanded proximal caudals. The premaxilla is fused and expanded and is a good indication of moving dirt with their snout. The arms are long, but no hands are known. *Oryctodromeus* was found in the bad lands of Montana and lived during the Mid-Cretaceous Blackleaf Formation, about 95 mya. Only skeletal fragments were known from that formation. The pubis/ischium is short, and body was 'short'. It is more similar to the Early Cretaceous *Zephyrosaurus* and Late Cretaceous *Orodromeus*. If having a short body is an indication of a fossorial life then perhaps *Parksosaurus* was a digger, unfortunately no premaxilla is known from *Parksosaurus*, so it isn't known if the premaxilla was fused. *Hypsilophodon* and other small ornithopods don't show fossorial skeletal features.

Unfortunately, the authors either ignore or didn't know about Bakker's article with *Drinker*. In that article he states that the long toes of *Drinker* were good for walking across marshes, possibly climbing (which I want to look into) and digging. Many specimens (undescribed unfortunately) have been found in an oval mudstone mass about a meter across. The perfect size for a burrow.

Also, not mentioned is Fastovsky et al, 1998 paper on *Protoceratops*. *Protoceratops* is an incredibly abundant dinosaur and several intact or nearly intact specimens have been found. Why? Surely they didn't stand or lie next to sand dunes and the dunes would fall on them. They could have been caught in deadly sandstorms, during flash floods -- or possibly in burrows. Some have been found 'standing up', others found in angles. The claws of *Protoceratops* are very broad and spade like, perfect for digging. Perhaps they dug burrows big enough for the males to use their frill to keep the burrow closed? Other dinosaurs that were possibly fossorial are *Psittacosaurus*, *Heterodontosaurus*, and *Mononykus* (ok, I've never actually believed it could dig). This isn't to say that only these dinosaurs were fossorial. I wouldn't doubt that there were more burrowing dinosaurs.

So, if you want to be creative, show a head popping out of the ground like a ground hog or prairie dog.

Another dinosaur that has spade like claws and short toes, is *Stegosaurus*. Surly there is no way *Stegosaurus* dug a burrow, but it certainly could and most likely did, dig. Either to get grubs, insects or tubers? Ceratopians also have spade like claws and probably also dug, possibly for nesting or finding food.

Bibliography

Bakker, R. T., 1990, A new Late Jurassic Vertebrate fauna, from the highest levels of the Morrison Formation at Como Bluff, Wyoming, With Comments on Morrison Biochronology. Part I: Biochronology: Hunteria v. 2, n. 6, p. 1-3.

Fastovsky, D. E., Watabe, M., and Badamgarav, D., 1998, Late Cretaceous dinosaur-bearing paleoenvironments, Gobi Desert, Mongolia: In: The Dinofest Symposium, Presented by. The Academy of Natural Sciences Philadelphia, Pennsylvania, April 17-19, 1998, edited by Wolberg, D. L., Gittis, K., Miller, S., Carey, L., and Raynor, A., p. 14-15.

Varricchio, D. J., Martin, A. J., and Katsura, Y., 2007, First trace and body fossils evidence of a burrowing, denning dinosaur: Proceedings of the Royal Society, Series B, Published on line, 8pp.

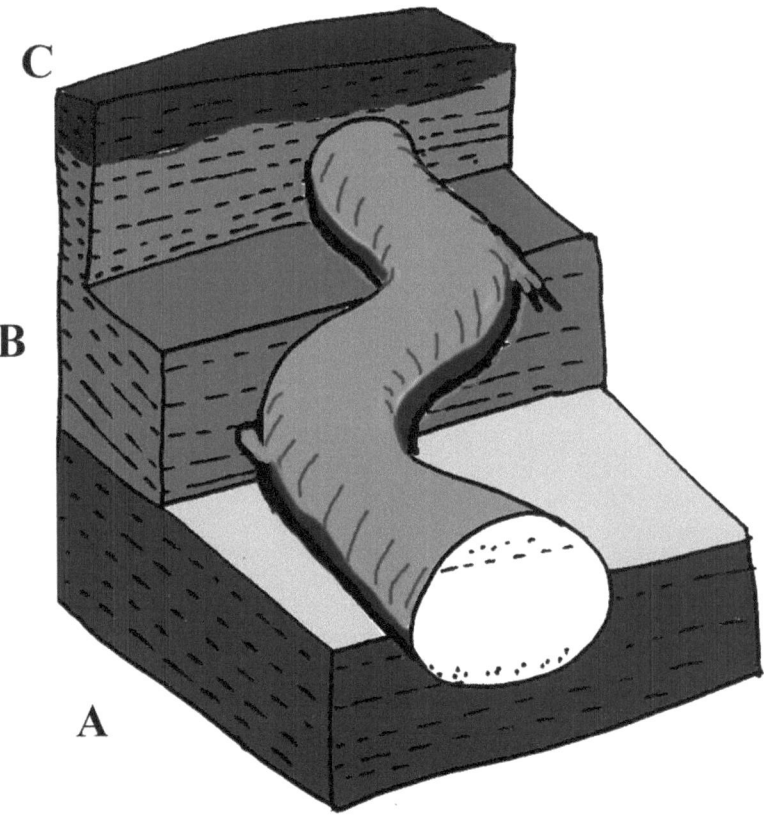

Figure 1): Burrow of *Oryctodromeus*, A) Mudstone, B) Claystone, C) Topsoil.

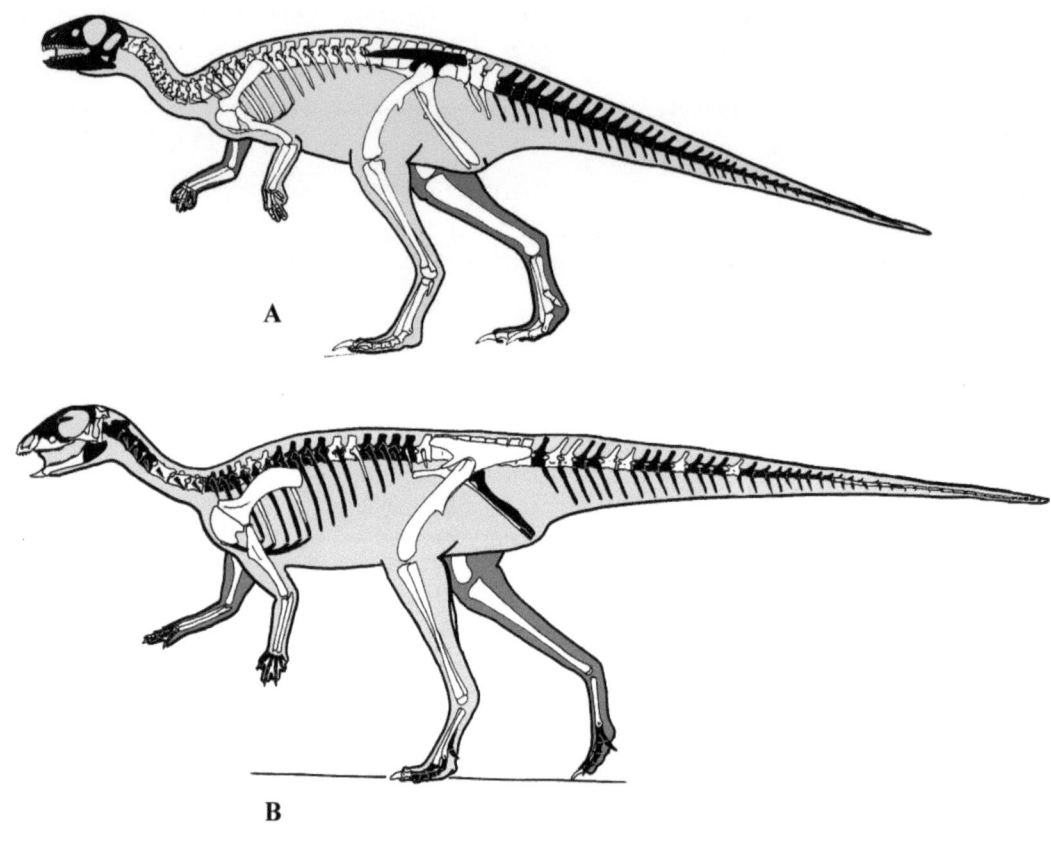

Figure 2): A) Skeleton of *Drinker*; B) Skeleton of *Oryctodromeus*.

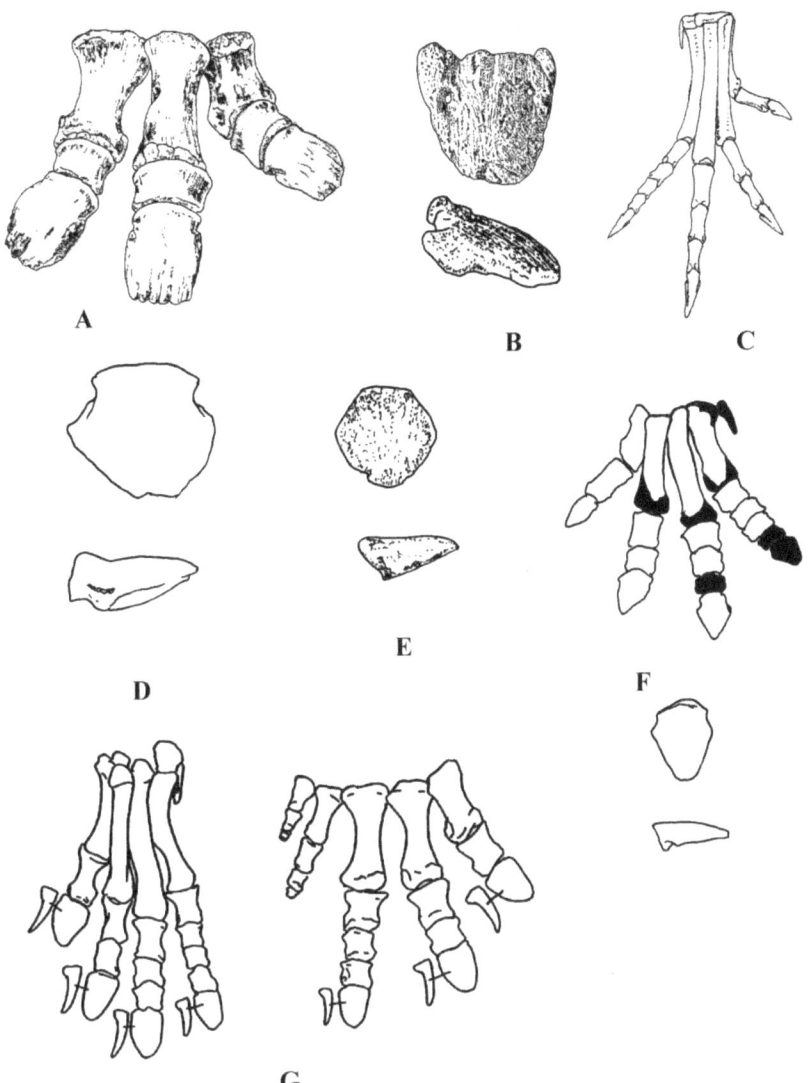

Figure 3): Claws and feet of digging dinosaurs; A) Pes of *Stegosaurus*; B) *Stegosaurus sulcatus* claw in dorsal and side view; C) Pes of *Drinker*; D) Manus claw of *Triceratops serratus* in dorsal and side view; E) Pes claw of *Triceratops horridus* in dorsal and side view; F) Pes of *Avaceratops* and claw in dorsal and side view; G) Manus and pes of *Protoceratops* with side view of claws.

53

Figure 4): *Protoceratops* digging.

Ford, T. L., 2007, How to Draw Dinosaurs, Neck muscles and feeding strategies in large theropods. Prehistoric Times, n. 83, p. 28-29.

Chapter 11

Neck muscles and feeding strategies in large theropods

I've been asked a few times to do an article on dinosaur muscles. I shied away because even with the dinosaur muscle bone scars that show where most of the muscles should be, there are still all those smaller ones. A new paper by Eric Snively and Anthony P. Russell on the neck muscles of large theropods was just published and I thought I'd write about it (Snively, E., and Russell, A. P., 2007 Functional variation of neck muscles and their relation to feeding style in Tyrannosauridae and other large theropod dinosaurs: The Anatomical Record, v. 290, p. 934-957).

The authors looked at tyrannosaurids, allosaurids, abelisaurids and ceratosaurids. The article is a bit technical on myology (the study of muscles and muscle tissue) so I will paraphrase what they wrote. There are several types of muscles, but we are only interested in skeletal muscles. The terminology for how muscles attach to a skeleton is origin (originate) and insert. The origin is where the muscle 'starts/begins' and it ends at its insertion and tapers into glistening white tendons which attaches the muscle to the bone. The insertion is pulled toward the origin of the muscle.

Anterior parietal

The parietal is a major muscle attachment area for both the jaw muscles (anterior side) and neck muscles (posterior side). The jaw muscles extend from dentary to various parts of the posterior end of the skull. The dorsal margin of the surangular has a ridge which some muscles attach tom the *m. adductor mandiubulae externus superficalis* (m. a. m. e. s.) originates on the upper ridge on the surangular on the outside and inserts in the upper inside of the skull. The *m. adductor mandibulae externus medialis* (m. a. m. e. m.) originates on the inside of the surangular/angular and inserts on the anterior parietal. The depth of the surangular/angular will help determine how much jaw muscles there was. In nearly all theropods only *Tyrannosaurus* has the largest surangular/angular to the point where it is nearly the same height as the skull itself, which gives the skull its huge look. *Tyrannosaurus* also has the largest, thickest teeth of any theropod. They were crushers, able to even break the ilium and horns of *Triceratops*! The upper back of the angular had the *m. depressor mandibulare* (m. d. m.) which is very small and thin inserted onto the paroccipital process, and the lower edge of the angular had the *m. pterygoideus posterior/ventralis* (m. pt. post) which wrapped around the angular from the outside to the inside.

Posterior parietal and skull

The *m. transversospinalis capitis* (m. t. c.) originated from the tips of cervical neural spines (beginning at cervical 9) and inserted on the dorsal edge of the parietal. Depending on the shape of the parietal determined how think this muscle was. In *Ceratosaurus* the parietal is small and so was the muscle. In *Allosaurus* the muscle was larger and in *Tyrannosaurus*, it is the largest. Underneath that muscle was the *m. splenius capitis* (m. s. c.) which originated from the anterolateral surface of the C2 neural spine and inserted on the posterior surface of the parietal.

The *m. complexus* (m. complexus) originated from tall cervical epipophyses and inserted on the posterior aspect of the squamosal. Inside of that is the *m. iliocostalis captis superficialis* (m. l. c. s.) which originated from the distal, lateral or ventrolateral portions of the transverse processes (beginning with C5 or C5, with the posterior-most origin being C10 or D1) and inserted on the lateral extremity of the paroccipital process. And under that is the *m. iliocostalis capitis* (m. i. c.) which originated from the fascia surrounding the cervical rib shafts and /or form the large proximal surfaces of posterior cervical ribs and inserts on the ventral portion of the paroccipital process or its ventral edge. The *m. longissimus capitis profundus* (m. l. c. p.) originated from the transverse processes of the anterior cervicals and inserted ventromedially on inserted to the basioccipital tuberosites.

The *m. rectus capitis ventralis* (m. r. c. v.) originated from the hypopophyses and ventral surfaces of the cervical centra and inserted on the occipital condyle. The *m. longus colli dorsalis/m. transversopinalis cervicis* (m. l. c. d./ t. cerv) originated on the lateral surfaces of the neural spines of C4-C10, ventral to the origins of m. t. c. and inserted onto posteriorly-facing scars of the anterior epipophyses and perhaps onto the dorsal surfaces of some posterior cervical epipophyses. What does all this mean?

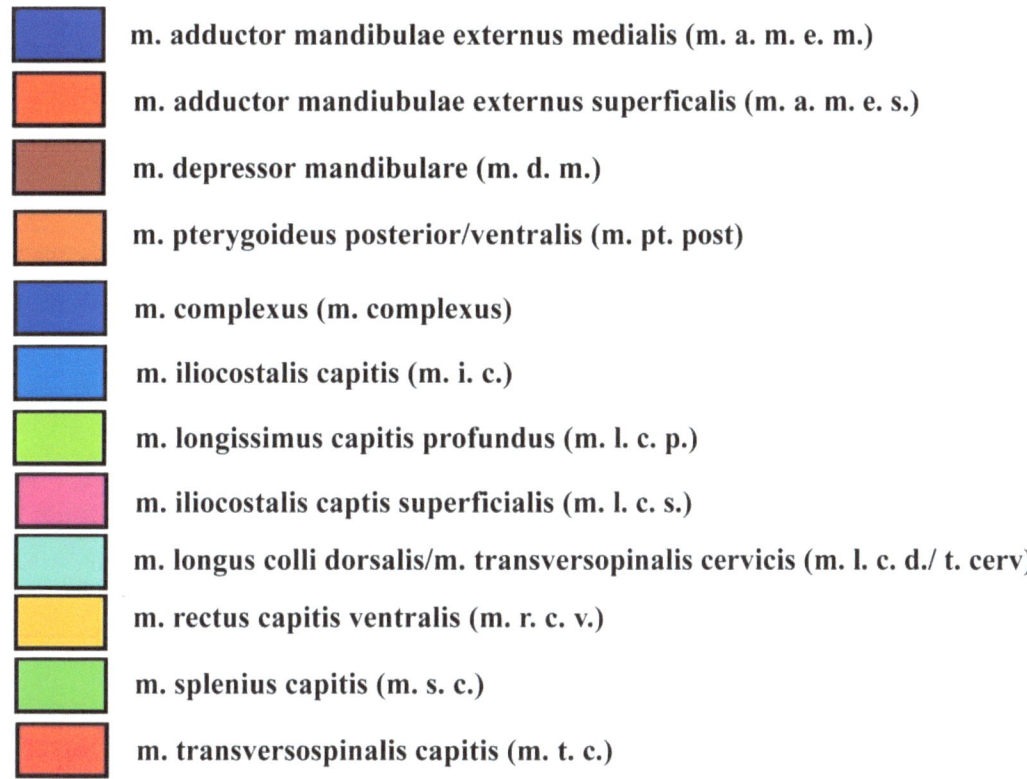

Figure 1). Color index of muscle (after Snively et al., 2007).

The shape of the parietal varies in theropods. They are tall, short or broad. The theropods with tall parietals (Abeliosaurids; *Carnotaurus*, *Majungathols*, Tyrannosaurids; *Gorgosaurus*, *Albertosaurus*, allosaurids; *Allosaurus*, *Sinoraptor*) and with broad parietals (*Tyrannosaurus*) had rugose scaring of the m. t. c suggesting a notably strong dorsiflexor. The rugosities indicate the muscles pulled with considerable force on the insertion and with the tall/broad parietal suggest a strong dorsifelxion by the m. t. c.; in other words, it was able to move its head upward with great strength. In *Ceratosaurus* and *Monolophosaurus* the parietals are short and small, and the m. t. c. would have been less effective.

The squamosal's are vertically extensive above the occipital condyle (*Ceratosaurus*, carnosaurs and tyrannosaurine tyrannosaurids) and the m. c. was very large. In conjunction of the m. t. c. the dorsiflexion would have been greatly augmented. Even though in *Ceratosaurus* and *Monolophosaurus* had a weaker m. t. c., the m. c. would have helped compensate for this weakness. In albertosaurine tyrannosaurids the squamosals are low and the m. c. would have not have been as strong as in tyrannosaurine tyrannosaurids, but the taller parietal may have had better leverage of the m. t. c. The morphology of the squamosal influences the capacity of m. c. for lateral flexion. The breath across the squamosal in tyrannosaurids (especially in *Tyrannosaurus rex*) suggests stronger leverage for lateral flexion than other theropods. In albertosaurine tyrannosaurids, the lateral movement was greater than the dorsiflexion.

The cranial ventroflexion entails straight forward inferences in large theropods, though they are not all the same in all theropods. *Ceratosaurus* has a massive basitubera with the large insertions of the m. l. c. p. and m. r. c. v. indicating that these muscles are very large.

Neural spine heights were also used in the study. It turns out that even when a scaled down *Tyrannosaurus rex* to the same size of *Gorgosaurus,* it had larger dorsiflexor by at least 6 times. In absolute terms, *T. rex* specimens had approximately 1.5 times as long as the *Gorgosaurus*, indicating the dorsiflexor muscles capable of exerting at least 9 times more force. Even though *T. rex* is a larger and bulkier animal it

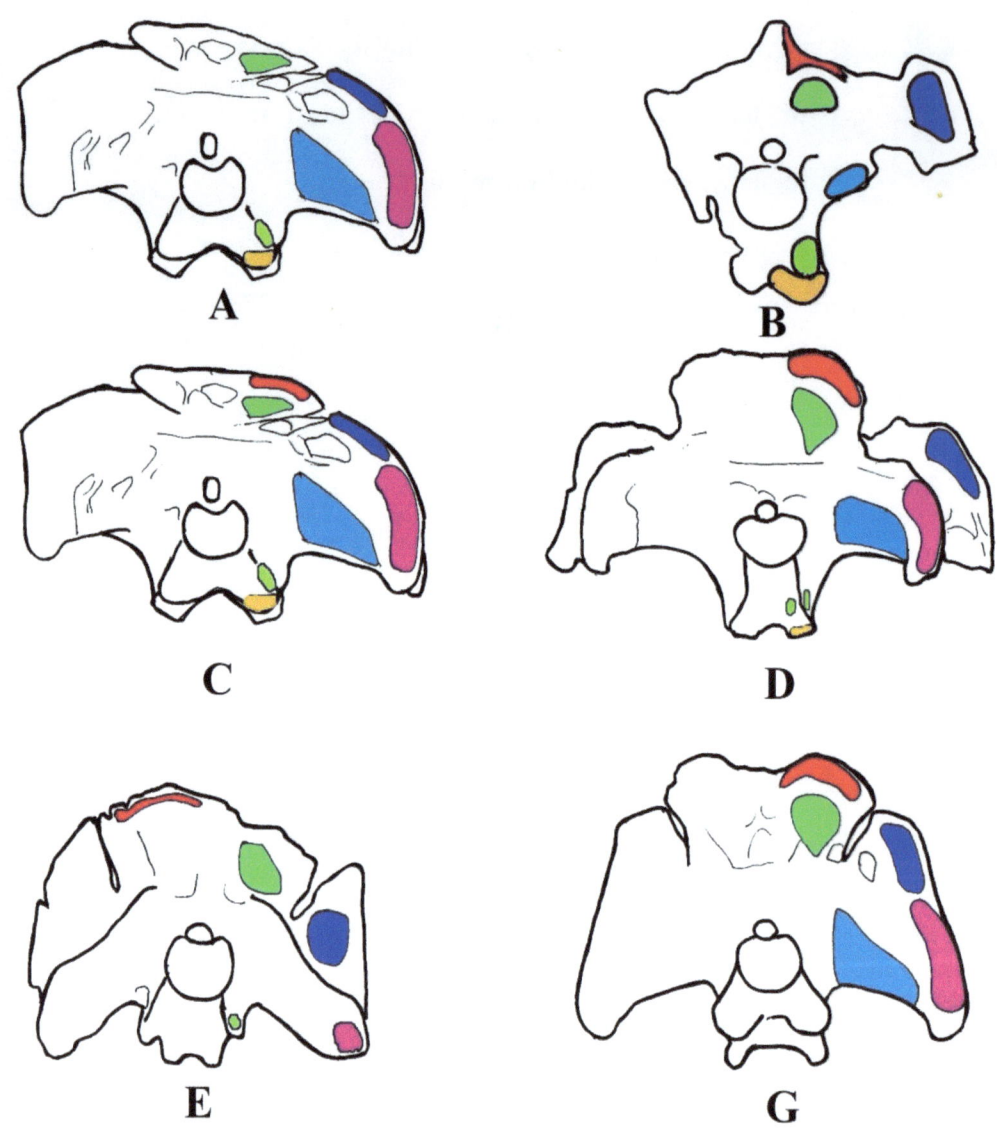

Figure 2) Posterior of theropod skulls; A) BHI 3033 (Stan) showing the missing portion of parietal; B) *Ceratosaurus* (BYU 881/12893); C) *Tyrannosaurus rex* (FMNH PR 2081, "Sue"); D) *Gorgosaurus* sp (CMI 2001.89.1); E *Allosaurus fragilis* (BYU 671/671/8901); F) *Daspletosaurus torosus* (CMN 8505) (after Snively et al., 2007).

did not lose its dorsoflexion ability for its size. Dorsiflexion (head moving up); All theropods had good dorsiflexion, though some are better than others. *Ceratosaurus* had the weakest dorsiflexion.

Ventroflexion (head moving down). Tyrannosaurids had the weakest ventroflexion, along with ablisaurids and *Monolophosaurus*. *Allosaurus* had better ventroflexion while *Ceratosaurus* had the greatest.

Lateroflexion (side to side). Tyrannosaurine tyrannosaurids had the greatest lateroflexion movement, though in albertosaurine tyrannosaurids, abelisaurids and *Ceratosaurus* had good lateroflexion. What does this say for theropod feeding tactics?

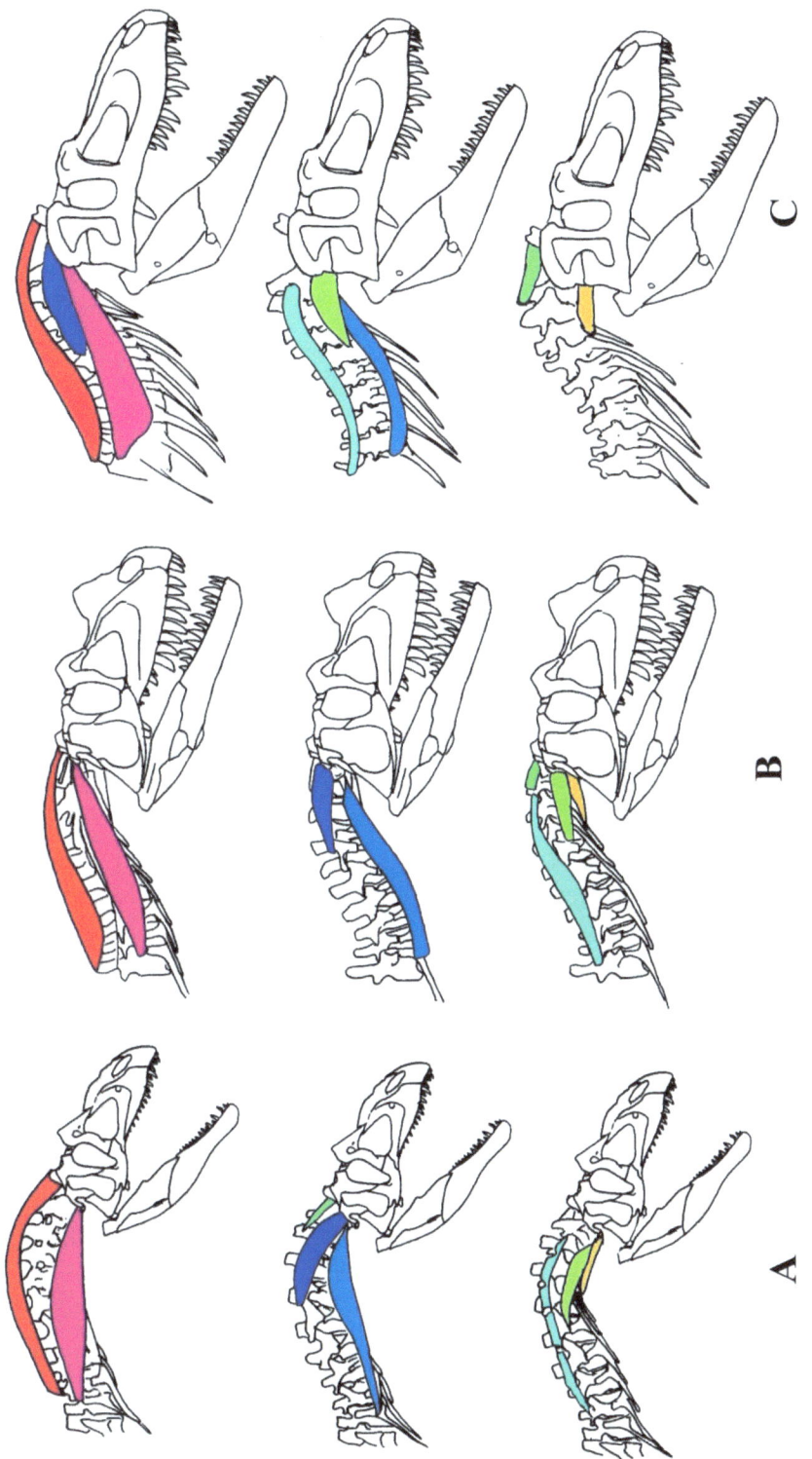

Figure 3) Side view of muscles of theropod skulls; A) *Allosaurus fragilis* (USNM 4734); B) *Ceratosaurus nasicornis* (USNM 4735); C) *Gorgosaurus* sp (CMI 2001.89.1) (after Snively et al., 2007).

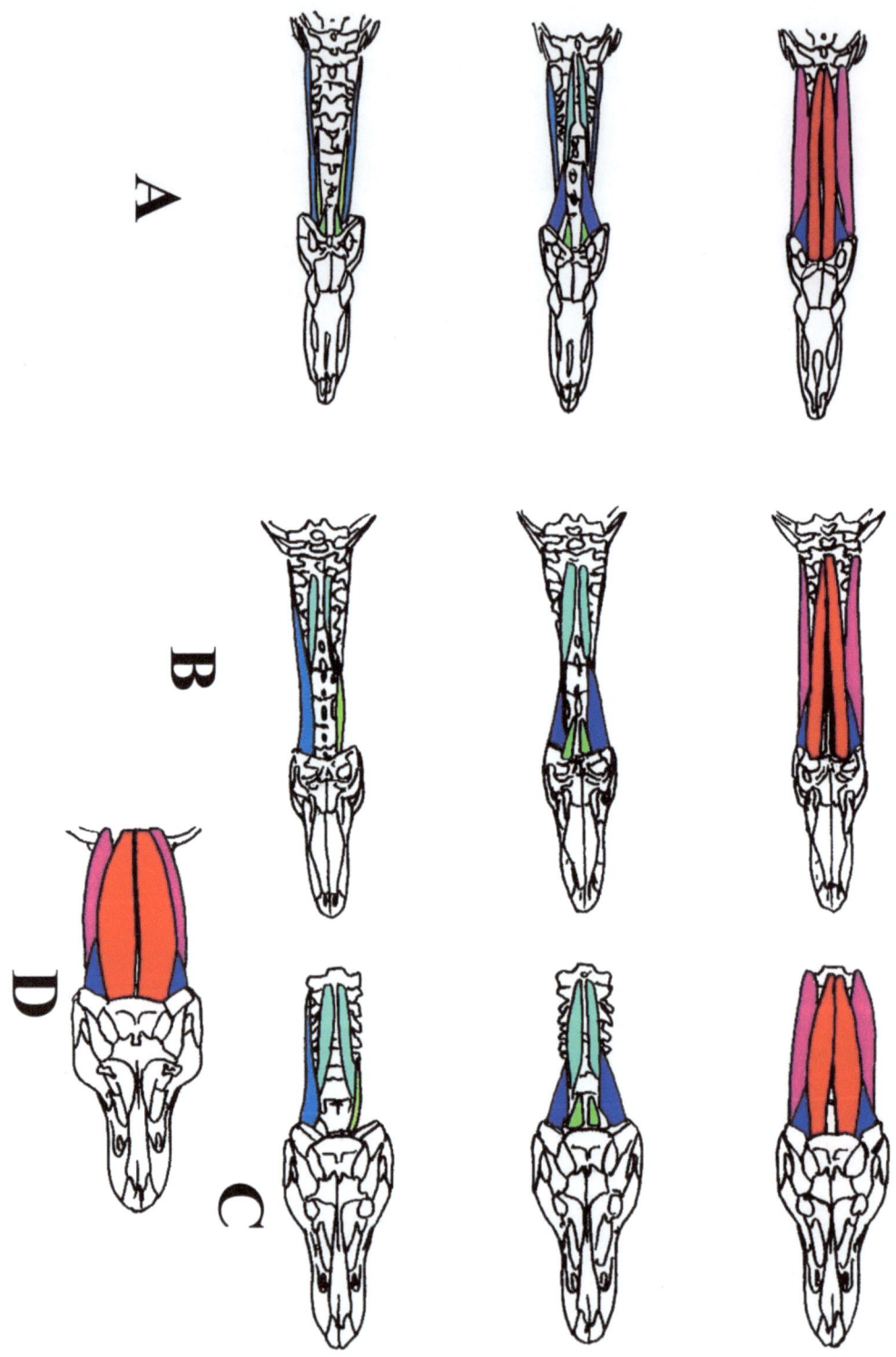

Figure 4) Top view of muscles; A) *Ceratosaurus nasicornis* (USNM 4735); B) *Allosaurus fragilis* (USNM 4734); C) Juvenile *Tyrannosaurus rex* (BMRP 2002.4.1); D) *Tyrannosaurus rex* (AMNH 5027) (after Snively et al., 2007).
Slice and rake feeders.

Its large ziphodont teeth and inferred large muscles for ventroflexion indicates *Ceratosaurus* was probably a slice and rake feeder. It had particularly rapid and powerful head dorsiflexion and lateroflexion, though not particularly forceful movements of its head. *Ceratosaurus* and Abelisaurids have short forelimbs and manual digits but have large cnemial crests on the tibia suggests powerful knee extensions during locomotion. In other words, they approached prey with rapid acceleration to relatively low speeds with little engagement with its forelimbs. *Ceratosaurus* with its large ziphdont teeth, lighter neck than abelisaurids and high geared craniocervical movement arms suggests that it used rapid slashing strikes with its jaws when engaging large prey and possibly raked its upper teeth through its prey. This is analogous with the Komodo monitor.

Strike and tear/pull feeders.

This kind of feeding would involve powerful ventroflexion to push its upper teeth through flesh and to strike downward at prey. *Allosaurus* and *Sinraptor* have powerful ventroflexive kinematics of their skulls indicates it used its upper teeth to bite into its prey. Its neck was extremely maneuverable and appears to have been capable of rapid and high-excursion movements of its head and neck. With its large forelimbs, large hands and claws *Allosaurus* was a formable predator capable of both gripping and raking through flesh and had multiple weapons at its disposal to subdue prey, both small and large.

Puncture and pull/shake feeders.

The idea here is that tyrannosaurid would puncture the soft tissues and bones of prey and pull to excise the tissues from the prey's body by lateral shaking and pulling dorsally rapidly. This strategy is not a new suggestion for tyrannosaurids and has been suggested before based on tooth marks and tooth morphology. Albertosaurine tyrannosaurids and other large theropods were not as adapt in lateral movements, but probably used this form of feeding like modern crocodilians when on land.

Tyrannosaurus rex had the most powerful neck and jaw muscles of any theropod. Its teeth are the largest capable of not only ripping through flesh but breaking large bones (such as the ilium and horns in *Triceratops*). There is a *Triceratops* femur that has multiple *T. rex* holes in it (Mike Triebold calls this its *Tyrannosaurus* biscuit). Even though it has relatively short forelimbs (the humerus is actually longer to its femur, 1/3 in length than in *Albertosaurus* and *Tarbosaurus* which is ¼ in length) it was more than capable to catch prey, rip out a huge chunk of flesh and bone by moving its head rapidly and forcefully dorsally and from side to side. It must have been awe inspiring site to see. To me this paper confirms *T. rex* was an active hunter and not a passive scavenger.

What this paper also tells me that my theory that Stan (BHI 3033) the *T. rex* died by starvation is now more probable. The parietal in Stan has a tooth marked hole and the upper left portion missing. It does show healing, so it lived for a while before it died. Also, some of the teeth haven't fallen out. I've talked to several paleontologists about this and they aren't convinced. They say the teeth have just moved during deformation/fossilization as seen in other fossils. This is incorrect because you can see the next tooth pushing out the one already there and this has been confirmed to me by Neal Larson. Not only would a chunk of the jaw muscles would be missing (*m. adductor mandibulae externus medialis* [m. a. m. e. m.]) but the entire insertion of the m. t. c. is would also be missing. The m. t. c. is greatly used in dorsoflexion and with that missing it would hinder its feeding abilities. Not being able to bite down or move its head would mean it would slowly starve to death, (let alone the shock to its system). It could close its mouth but not with enough strength needed to capture prey. Editor's note: Since the writing of this paper, I have also noted that Stan is missing portions of its cervical vertebrae, on the same side of the bitten off parietal, which also indicates some of its dorsal cervical neck vertebrae was also missing (m.t.c).

Bibliography

Snively, E., and Russell, A. P., 2007, Functional variation of neck muscles and their relation to feeding style in Tyrannosauridae and other large theropod dinosaurs: The Anatomical Record, v. 290, p. 934-957.

Snively, E., and Russell, A. P., 2007, Functional morphology of neck musculature in the Tyrannosauridae (Dinosauria, Theropoda) as determined via a hierarchical inferential approach: Zoological Journal of the Linnean Society, v. 151, p. 759-808.

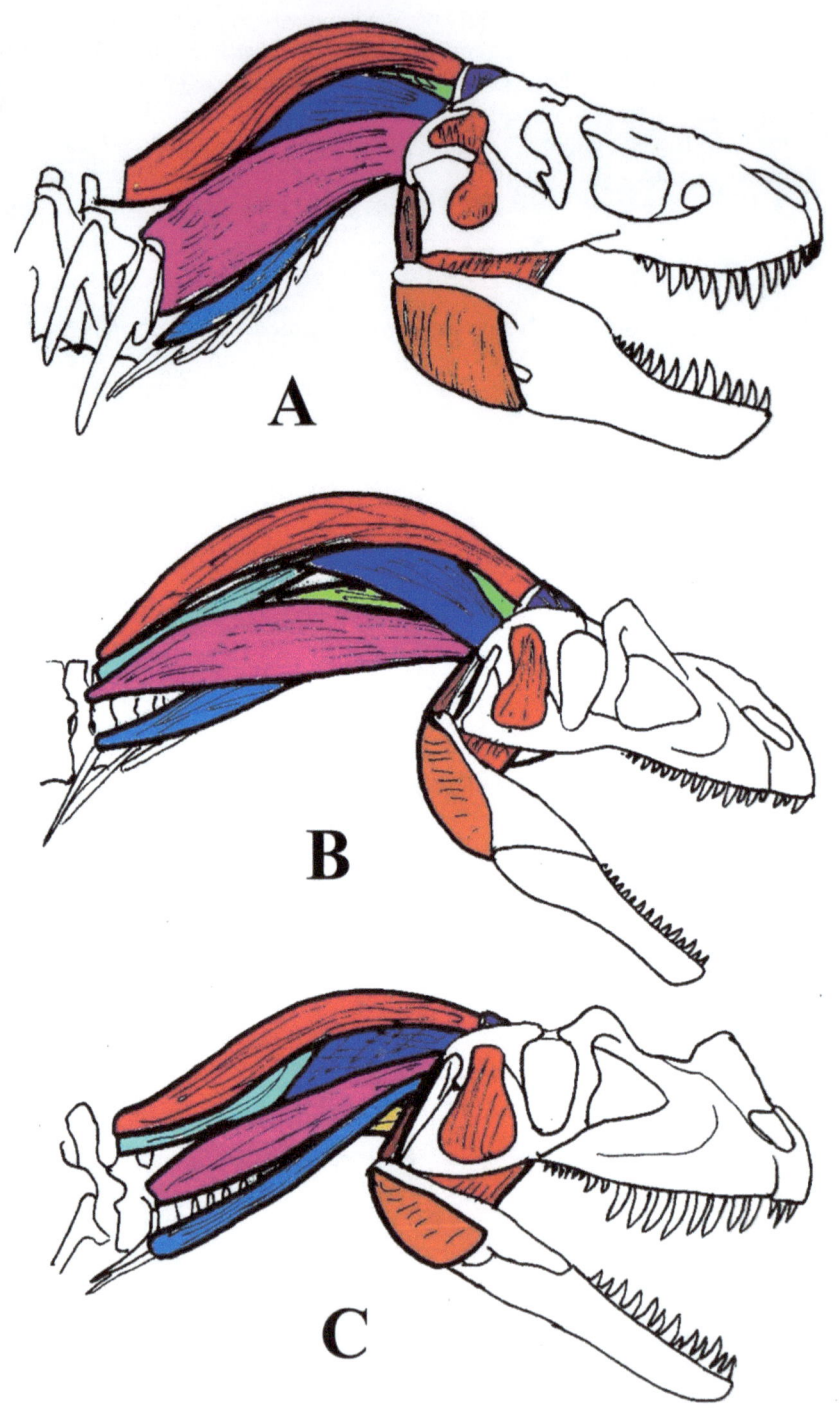

Figure 5) Side view of all the theropod muscles; A) *Tyrannosaurus rex;* B) *Allosaurus fragilis*; C) *Ceratosaurus nasicornis* (after Snively et al., 2007).

62

Ford, T. L., 2008, How to Draw Dinosaurs, And Now, Back to Sauropods! Prehistoric Times, n. 84, p. 18-19.

Chapter 12

And now, Back to Sauropods

This article stems from a query from a paleoartist. He wanted to know where he could find illustrations of sauropod dorsal vertebrae that he was interested in. There really isn't one place for him to find illustrations of the vertebrae he was interested in and I thought that would make a good HTDD article. He asked about *Apatosaurus, Diplodocus, Barosaurus, Amargasaurus, Suuwassea, Rebbachisaurus,* and *Dicreaosaurus*. I will also be talking about *Limaysaurus, Brachiosaurus, Giraffatitan* and *Brontosaurus*. And yes, I do believe *Giraffatitan* and *Brontosaurus* are valid genera. The problem is some of the sauropods he was asking about are only known from fragmentary skeletons with little dorsal vertebrae preserved.

The dorsal vertebrae of *Brachiosaurus* and *Giraffatitan* are very different from each other and I believe Greg Paul was right in separating the two (though he made them subspecies of *Brachiosaurus*, and it was George Olshevsky who finally split the two). Brachiosaurids is believed to have had 12 dorsal vertebrae (so far there is no complete brachiosaurid known) which is two more dorsal vertebrae than what diplodocids have. The neural spines are not split. The holotype of *Brachiosaurus altithorax* has the 6th to the 12th dorsal vertebrae preserved while the holotype of *Giraffatitan* (*Brachiosaurus*) *brancai* has the 4th, 8th, and the 10th to the 12th. The dorsal vertebrae in *Brachiosaurus* are larger and stouter than in *Giraffatitan*; the neural spines are wider, and the centrum is larger. To me these differences show that they are different genera.

Diplodocimorpha has two superfamilies, Rebbachisauroidea and Diplodocoidea. Rebbachisauroidea has only one family, the Rebbachisauridae, and the Diplodocoidea has Diplodocoidae and Dicraeosauridae. Both have 10 dorsal vertebrae, though *Barosaurus* has 9. The dorsal vertebrae in Rebbachisauroidea are all single spined (or not split) and Diplodocoidea are split from dorsal 1 to dorsal 6. Though the genera are similar they can be easily distinguished from one another.

Rebbachisaurids have tall neural spines. There are two species of *Rebbachisaurus*, *R. garasbae* and *R. tamesnensis*; there are fewer specimens known of the former. *Rebbachisaurus garasbae* is known from a fragmentary specimen. Only a few dorsal vertebrae are known along with a scapula, humerus, and sacrum. When the specimen was first described by Lovacat he reported that the dorsal neural spines were very tall, but he never figured it. It wasn't until Bonaparte (1999) that the dorsal vertebra was figured. The top of the neural spine is missing so the actual height isn't known. The vertebra is a meter and a half tall. A third species was named by Calvo and Salgado (1995) as *R. tessonei*. But later it was shown to be a separate genus, *Limaysaurus tessonei* by Salgado, Garrido, Cocca & Cicca, (2004). It is a more complete specimen, but its neural spines aren't as tall as *Rebbachisaurus*.

In diplodocoids. the first dorsal neural spines are short and get taller toward the sacrum. *Suuwassea* has three fragmentary dorsal vertebrae; 1st, 2nd and 3rd. The neural spines are split and are the shortest known in diplodocids. The holotype of *Apatosaurus ajax* has dorsals 1 to 4, and 6 to 10, and the referred species of *Atlantosaurus immanis* has dorsals 1, 3, 5, 8 and fragmentary 6, 7, 9, 10. Unfortunately only a few of the dorsal vertebrae have been figured of the type (that I know of). What has been figured shows that the centra are thinner (front to back) and the neural spines are short. A new paper by Upchurch, Tomida and Barrett (2004) described a new specimen of *Apatosaurus ajax*. It has a complete dorsal series but they only figure 4 of them. The type of *Apatosaurus ajax* is the largest of the *Apatosaurus* and *Brontosaurus* species. The Dinosaur Journey Museum in Fruita Colorado has a huge sacrum and other skeletal elements of what I believe is an adult *Apatosaurus*.

Gilmore wrote a monograph about *Apatosaurus louisae* and *excelsus*. I believe there are differences between *Brontosaurus* and *Apatosaurus*, and *Brontosaurus* is a valid genus. The neural spines are taller and centra wider in *Brontosaurus*. That is why I'm showing all three species. The last dorsal neural spines are the tallest of the diplodocids.

Barosaurus lentus has short neural spines. They are taller than *Suwassia*, but shorter than other diplodocids. It's interesting that the cervical vertebrae of *Barosaurus* are one third longer than *Diplodocus*, but the caudal vertebrae are one third shorter so overall the animals are about the length. This ratio is about the same for *Apatosaurus* and *Brontosaurus* and is one of the reasons I believe they are distinct.

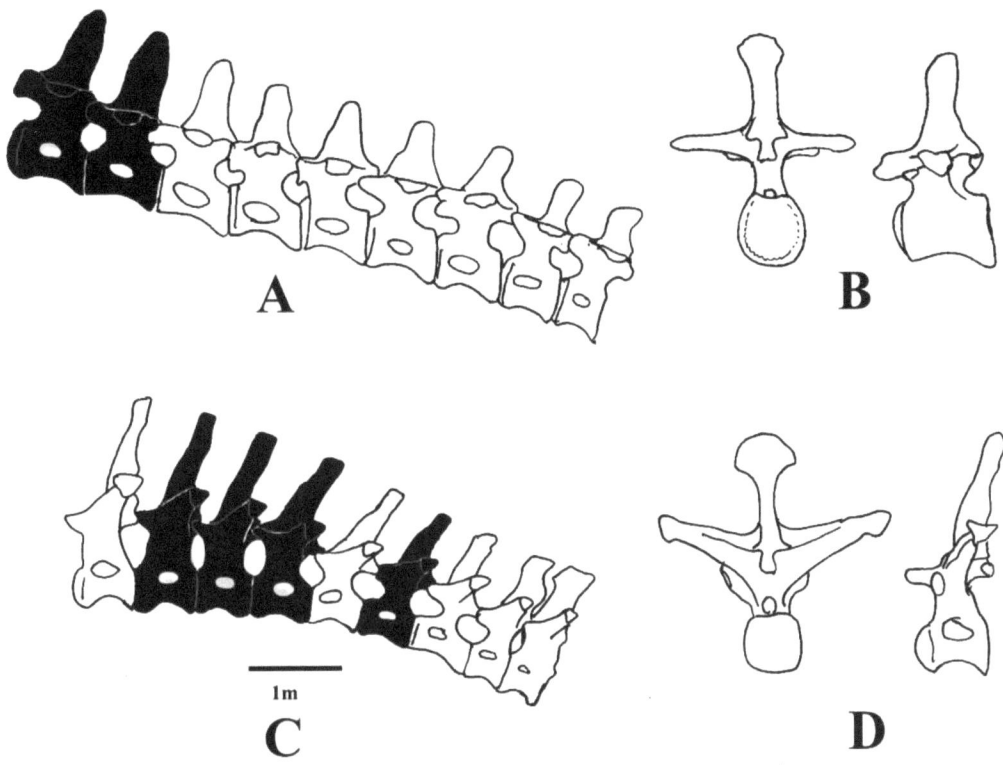

Figure 1) A) Side view of *Brachiosaurus altithorax* dorsal vertebrae from the 4th to 12th; B) Side and anterior view of dorsal 4; C) Side view of *Giraffatitan branchi* dorsal vertebrae from 4th to 12th; D) Side and anterior view of dorsal 4 (modified from Greg Paul).

Diplodocus carnegii has slightly taller neural spines.

Dicraeosaurids have tall neural spines with *Amargasaurus* having the tallest. Like the other diplodocids the 1st dorsal to the 6th is split. *Amargasaurus* neural spines are more rod-like and may have either had a double sale or just the neural spines sticking out of the back.

Bibliography

Bonaparte, J. F., 1999, Evolucion de las vertebras presacras en Sauropodomorpha: Ameghinana, tomo 36, n. 2, p. 115-188.

Calvo, J. O., and Salgado, L., 1995, *Rebbachisaurus tessonei* sp. nov. A new sauropod from the Albian-Cenomanian of Argentina; New evidence of the origin of the Diplodocidae: Gaia, n. 11, p. 13-33.

Gilmore, C. W., 1936, Osteology of *Apatosaurus*, with special reference to specimens in the Carnegie Museum: Memories of the Carnegie Museum, v. 11, n. 4, p. 175-300.

Harris, J. D., 2006, Cranial osteology of *Suuwassea emilieae* (Sauropoda: Diplodocoidea: Flagellicaudata) from the Upper Jurassic Morrison Formation of Montana, USA: Journal of Vertebrate Paleontology, v. 26, n. 1, p. 88-102.

Janensch, W., 1929, Die Wirbelsaule der Gattung *Dicraeosaurus*: Palaeontographica Supplment, n. (1) teil 2, lief. 1, p. 37-133.

Lull, R. S., 1919, The Sauropod Dinosaur *Barosaurus* MARSH: Memoirs of the Connecticut Academy of Arts and Sciences, v. 6, p. 5-42.

Olshevsky, G., 1991, A Revison of the Parainfraclass Archosauria Cope, 1869, Excluding the Advanced Crocodyila. Mesozoic Menanderings #2 (1st printing): iv + 196pp.

Paul, G. S., 1988, The Brachiosaur giants of the Morrison and Tendaguru with a description of a new subgenus, *Giraffatitan*, and a comparison of the world's largest Dinosaurs: Hunteria, v. 2, n. 3, p. 1-

14.
Salgado, L., and Bonaparte, J. F., 1991, Un nuevo Sauropodo Dicraeosauridae, *Amaragasaurus cazui* gen. et sp. nov. De la Formacion La Amarga, Neocomiano de la Provincia del Neuquen, Argentina: Ameghiniana, v. 28, n. 3-4, p. 333-346.

Salgado, L., Garrido, A., Cocca, S. E., and Cocca, J. R., 2004, Lower Cretaceous rebbachisaurid sauropods from Cerro Aguada del Leon (Lohan Cura Formation), Neuquen Province, northwestern Patagonia, Argentina: Journal of Vertebrate Paleontology, v. 24, n. 4, p. 903-912.

Upchurch, P., Tomida, Y., and Barrett, P. M., 2004, A new specimen of *Apatosaurus ajax* (Sauropoda: Diplodocidae) from the Morrison Formation (Upper Jurassic) of Wyoming, USA: National Science Museum Monographs, n. 26, 108pp.

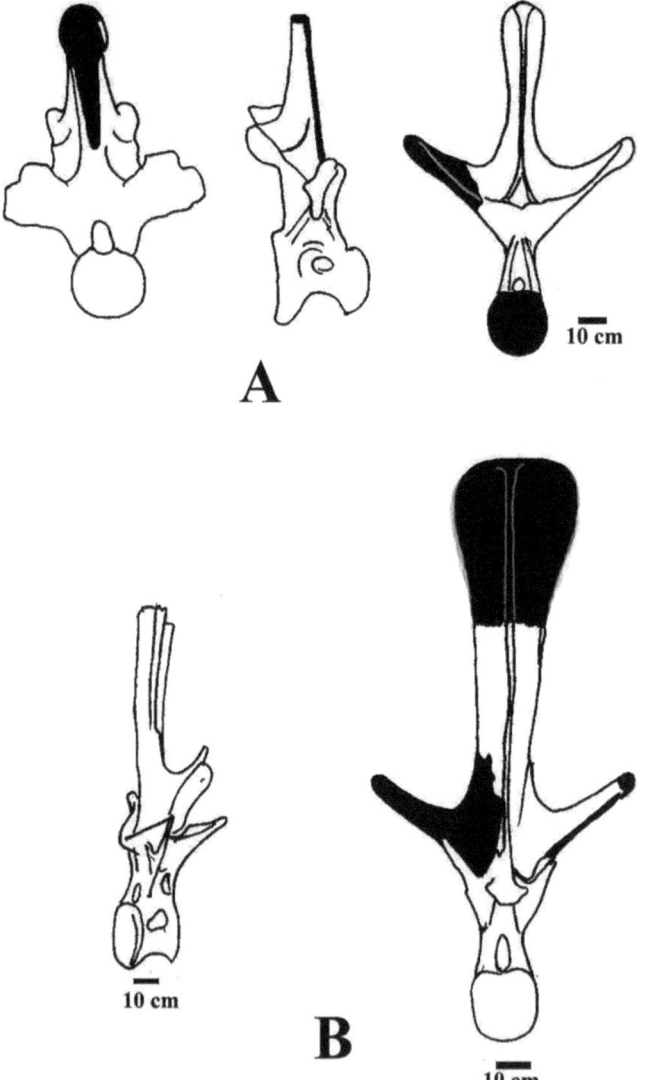

Figure 2) Rebbachisaurids. A) *Limaysaurus tessonei,* anterior and side view of a cervical vertebra (left and middle) and posterior mid-posterior dorsal vertebra; B) *Rebbachisaurus garasbae*, lateralposterior view and posterior view (The neural spine could be shorter than I have it).

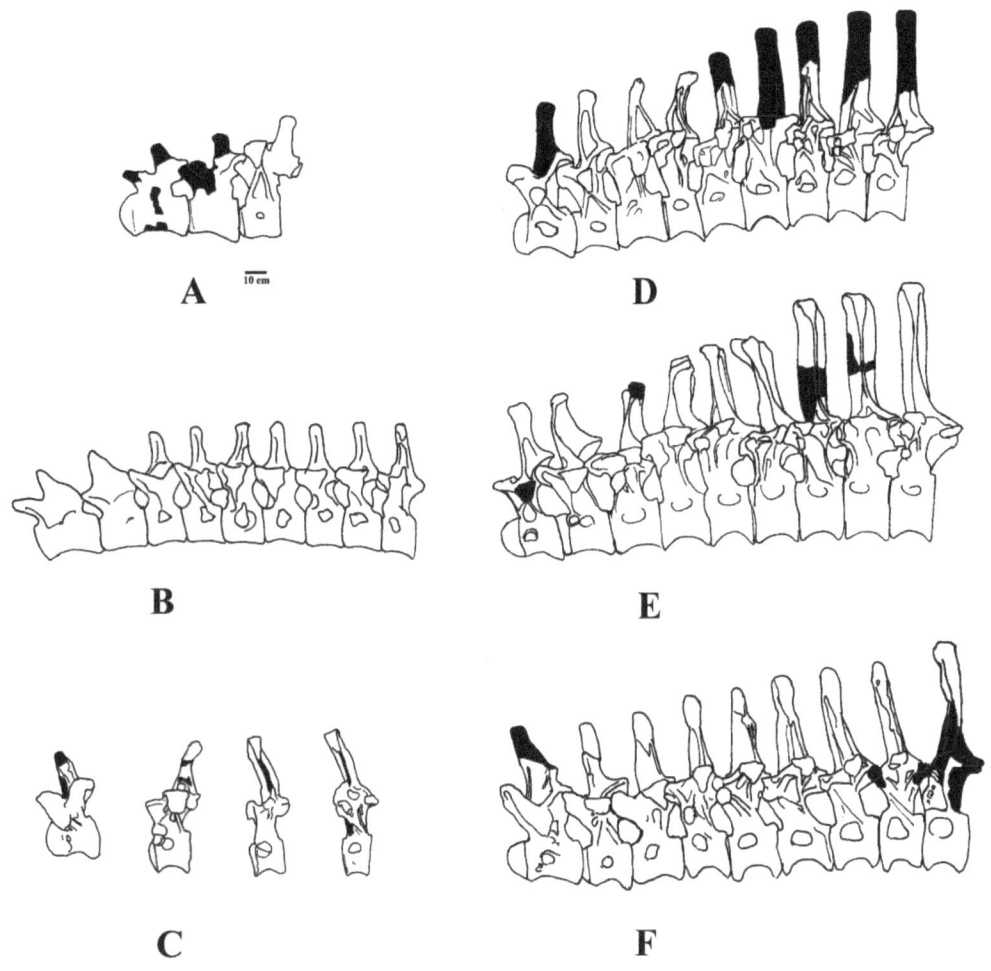

Figure 3) Side view of diplodocid vertebrae. A) *Suuwassea*; B) *Barosaurus*; C) *Apatosaurus ajax* (from Upchurch et al); D) *Brontosaurus excelsus*; E) *Brontosaurus louisae*; D) *Diplodocus*.

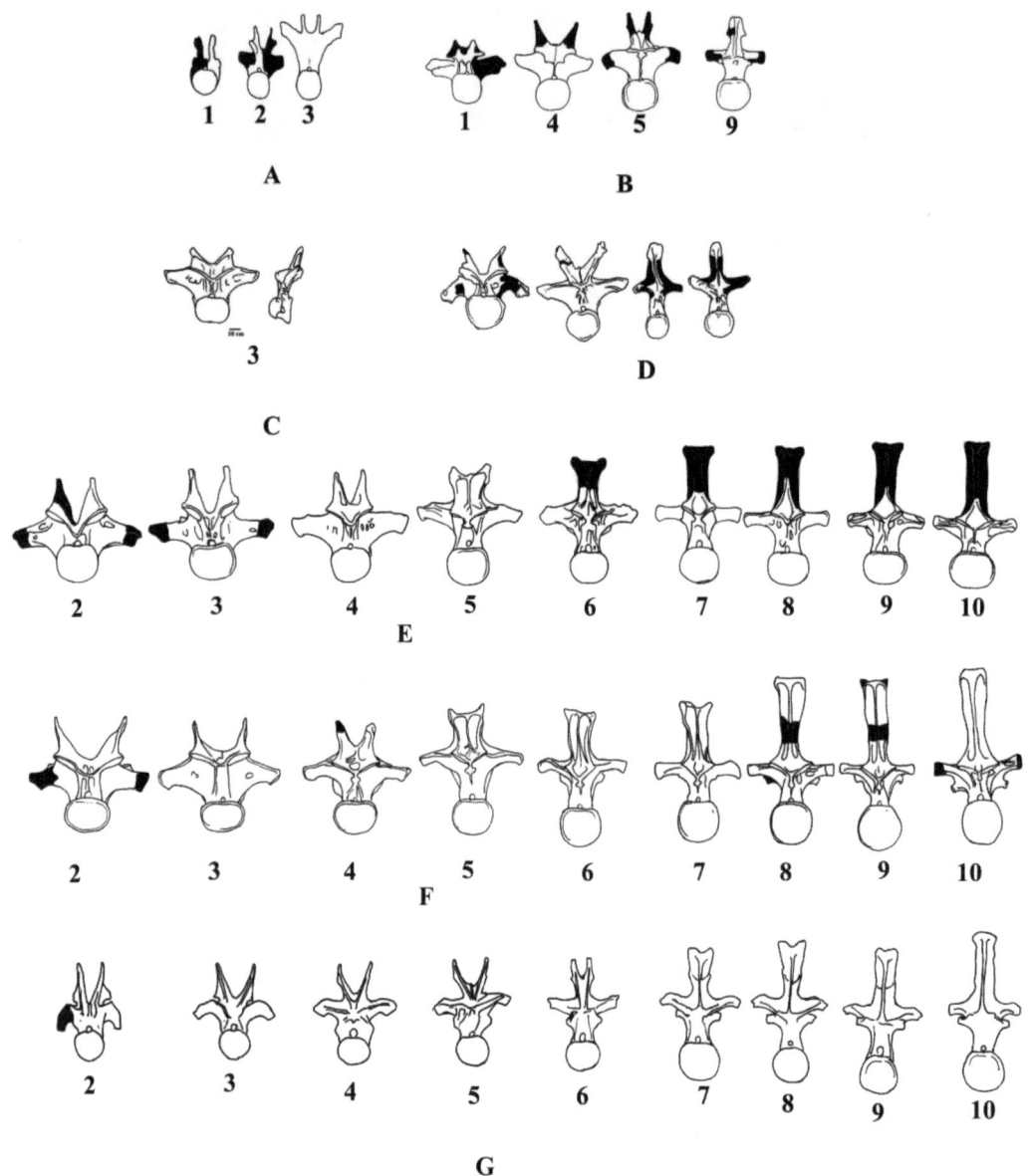

Figure 4) Posterior view of diplodocid vertebrae with vertebrae numbered. A) *Suuwassea*; B) *Barosaurus*; C) *Apatosaurus ajax* (type); D) *Apatosaurus ajax* (from Upchurch et al); E) *Brontosaurus excelsus*; F) *Brontosaurus louisae*; G) *Diplodocus*.

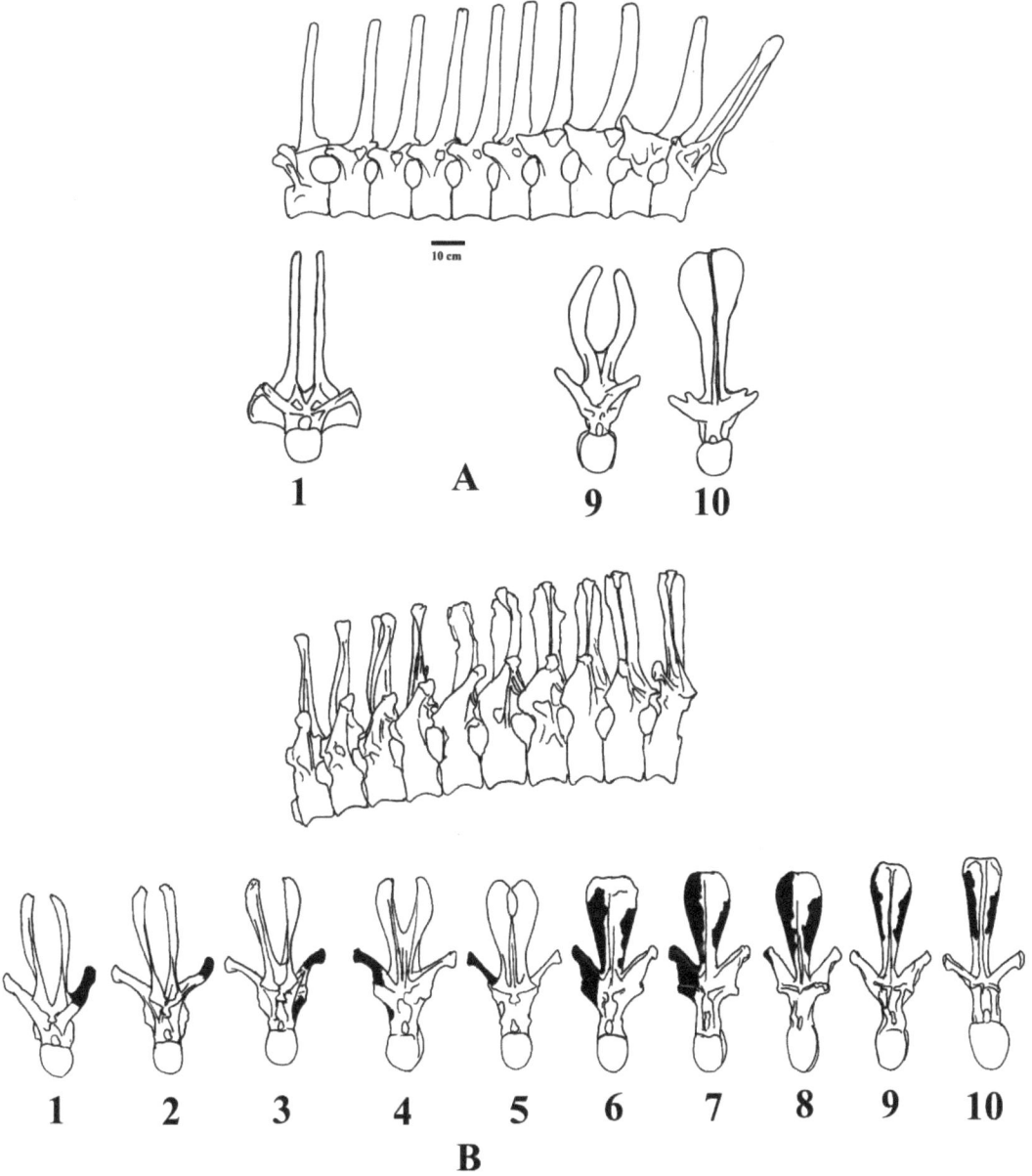

Figure 5) Dicraeosaurid dorsal vertebrae in side and posterior view; A) *Amargasaurus*; B) *Dicraeosaurus*.

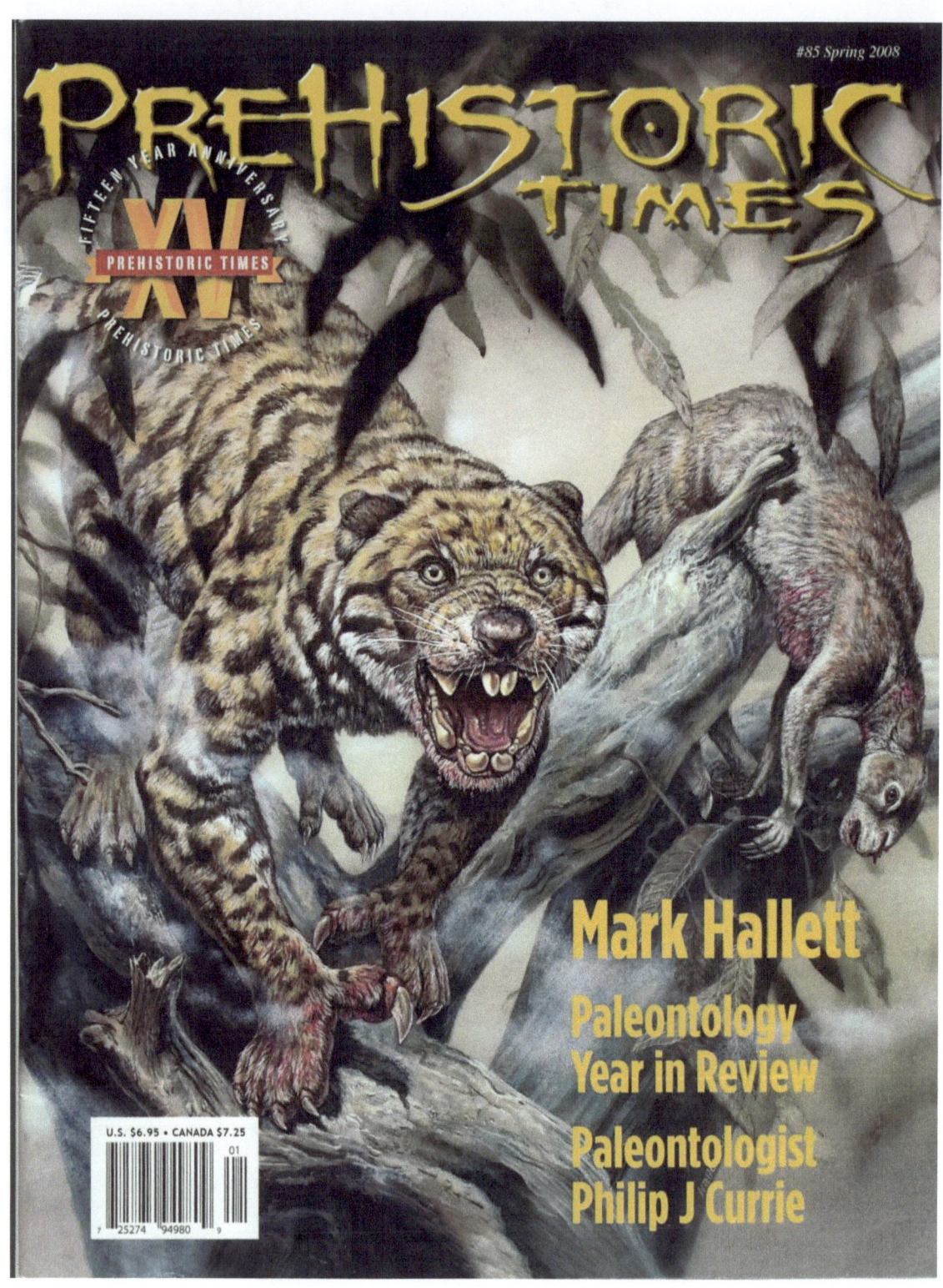

Ford, T. L., 2008, How to Draw Dinosaurs, Hadrosaurs revisited, Prehistoric Times, n. 85, p. 18-19.

Chapter 13

Hadrosaurs revisited

With this issue's featured dinosaur being *Edmontosaurus* I thought I'd go over my thoughts on *Edmontosaurus*/*Anatosaurus*/*Anatotitan* again and briefly go over Phillip Manning's recent documentary and books on a new mummified hadrosaur from North Dakota. For the last few decades *Edmontosaurus* has been the sole Late Cretaceous hadrosaurid hadrosaur from North America. Basically, three hadrosaurs were made into one; i.e. *Anatosaurus* and *Anatotitan* were sunk into *Edmontosaurus*. For nearly a decade I've thought and have written in PT that the three are valid genera. I believe there are differences in the skull and some of the skeleton that warrants this. Cladistically though there was only Maastrichtian hadrosaurid hadrosaur, *Edmontosaurus*. The oldest valid name of North American Maastrichtian hadrosaurid hadrosaurs is *Edmontosaurus* which was first coined by Lambe in 1917. *Anatosaurus* (*A. annectens*) was coined by Lull and Wright in 1942 and *Anatotitan* (*A. copei*) by Brett-Surman vide Chapman and Brett-Surman, 1990. All three are known from skulls and skeletons and *annectens* is known from several mummified specimens.

The skull of *Edmontosaurus* is shorter and higher than *Anatosaurus* and more so than *Anatotitan*. The skull of *Anatotitan* is lower and longer because of the lengthening of the muzzle via the premaxilla, nasal and dentary. The type skull of *Edmontosaurus* (CMN 2288) hasn't been completely prepared in the orbit area and the first illustrations of the skull were done conservatively. This is one of the things that bugs me, illustrating a skull conservatively and not following what the actual skull indicates. Sometimes there is crushing to some degree and an illustration correcting this is important. The Chicago skull of *Edmontosaurus* (CNHM P15003) is completely prepared and it shows that above the orbits there was a tall thin ridge. This ridge is usually either illustrated or reconstructed shorter than it actually was. The bill is the largest of any hadrosaur and behind the orbit the postorbital flares outward (this is being researched by Peter Larson). *Anatosaurus's* skull is more conservative in that the bill is smaller and the ridge over the orbit is nearly absent.

As I mentioned in a previous PT article the type skull of *Anatotitan* is dorso-ventrally crushed. I've tried to reconstruct the skull uncrushed. A few years ago, at the 2005 Tucson Rock/Fossil show there were two large, beautifully prepared skulls of Maastrichtian North American hadrosaurs. The first one I saw I thought it was an *Anatotitan*, but I was wrong because at another vender there had a skull of *Anatotitan* and the first skull was actually *Anatosaurus*. The skull of the *Anatotitan* was uncrushed and I was glad to see that I came close to getting the dimensions correct.

When drawn to the same scale the differences are easily seen between the genera. When drawn to the same length the skull of *Edmontosaurus* is the highest and *Anatotitan* is the lowest, but when drawn to the same height *Edmontosaurus* has the shortest skull and *Anatotitan* has the longest (Figures 1 and 2). *Edmontosaurus* is found in the Canadian Scollard, Horseshoe Canyon, Frenchman and possibly the Hell Creek Formations and hasn't been found with *Anatosaurus* and *Anatotitan*. *Anatosaurus* and *Anatotitan* on the other hand do overlap with each other with *Anatosaurus* being the more abundant.

Edmontosaurus regalis; Scollard Formation, Horseshoe Canyon Formation, Hell Creek Formation (or *Anatotitan*). Alberta Province, Saskatchewan Province, Canada.

Edmontosaurus sp; Horseshoe Canyon Formation, St. Mary River Formation. Alberta Province, Canada.

Anatosaurus annetens: Scollard Formation (= *Thespesius edmontonensis*), Hell Creek Formation, Lance Formation. Alberta Province, Canada, Montana, North Dakota, South Dakota, Wyoming.

Anatosaurus saskatchewanensis; Lance Formation. Saskatchewan Province, Canada.

Anatosaurus sp; Laramie Formation, Prince Creek Formation, Hell Creek Formation, Lance Formation.

Anatotitan copei; Hell Creek Formation, Lance Formation. North Dakota, South Dakota.

Genus species	Scollard Formation	Horseshoe Canyon Formation	Frenchman Formation	Hell Creek Formation	Lance Formation	St. Mary River Formation.
Edmontosaurus regalis	Yes	Yes	Yes	? or *Anatotitan*		
Edmontosaurus sp		Yes				Yes
Anatosaurus annectens	= *Thespesius edmontonensis*			Yes	Yes	
Anatosaurus sp				Yes	Yes	
Anatotitan copei				Yes ? or *Edmontosaurus*	Yes	

The argument made by Manning was about the reconstruction of '*Edmontosaurus*' tail was thicker than is usually reconstructed. The specimen he based this on is nicknamed Dakota and was first found in 1999 by a young and up and coming paleontologist, Tylor Lyson. The specimen he had found was a nearly complete mummified skeleton. They aren't sure quite yet how much of the skeleton there is because more prep work is needed and prepping it without destroying the skin is difficult. What they have found so far is that the tail was much broader than is usually illustrated. The tail muscles and skin extend further out from the body than is usually reconstructed giving the tail a much thicker look.

There have been several hadrosaur and lambeosaur mummies found throughout the decades. I've written about several of them in my articles in PT including the Senkenburg *Anatosaurus*, the MOR tail section (MOR V 007) (Figure 3), and Leonardo the *Brachylophosaurus*. Manning does mention two of them but not the MOR tail section. The mummified MOR tail section (part at the Museum of Rockies and the counterpart is at the Czerkas Dinosaur Museum in Blanding Utah) is from the mid section of the tail and shows that they had a thick tail with a serrated dorsal 'frill'. This 'frill' is what has prompted many artists to put the serrated edge on the back and tail on hadrosaurs.

The only illustration in his book is on the cover and that is also incorrect. At an earlier Tucson show a vender had a 'mummified' Maastrictian hadrosaur. They only had the skull, neck and tail on display. It was being sold as an *Edmontosaurus,* but I believe it to be *Anatosaurus*. The skull was still in the matrix and shows that it had a large gullar/ throat pouch. Stephen Czerkas and Greg Paul have argued that hadrosaurs had a thick horse-like neck and not the thin goosenecks that is usually depicted. The dorsal vertebrae had a steep downward turn around the midsection of the body, and the cervicals had a 'U'. The head was held at a right angle to the cervical vertebrae and this is true for both lambeosaurs and hadrosaurs. Even though the skeletal neck has a 'U' shape and was thin the skin and muscles on the mummified specimens indicate the neck was much thicker and horse-like.

As far as the tail is concerned where the chevrons stop does not indicate the end of the actual tail. It extended for at least another 10 feet or more and is whiplash-like. But this isn't to say that all hadrosaurs had this extra whiplash tail. There is the one that was at the Tucson Rock/Fossil show that I just mentioned, a juvenile at the Malta Museum (gen. sp indet.) and several more that are being studied by Canadian palaeontologist. Even though the tail at the Tucson show stopped ten feet after the last chevron it may not have been the actual end of the tail because these vertebrae were large and may have extended even further in life. It is interesting to note that when Rozhdestvensky published on *Saurolophus anustirostris* (PIN 551-8?) in 1957 he drew the skeleton with a longer than normal tail. The drawing was incorrect that it was dragging its tail but shows that if correct, it had a long whiplash like tail. The type species of *Saurolophus*, *S. osborni* (AMNH 5220) is a nearly complete specimen with the tip of the tail missing. The last 4 vertebrae lack chevrons and it was estimated by Brown in 1913 that there was probably 4 more feet of the tail was missing. If like *Anatosaurus* there would have been at least 10 more feet of the whiplash-tail. It may be prudent for paleontologist to look over more of the hadrosaur skeletons to see if they also had the whiplash-like tail.

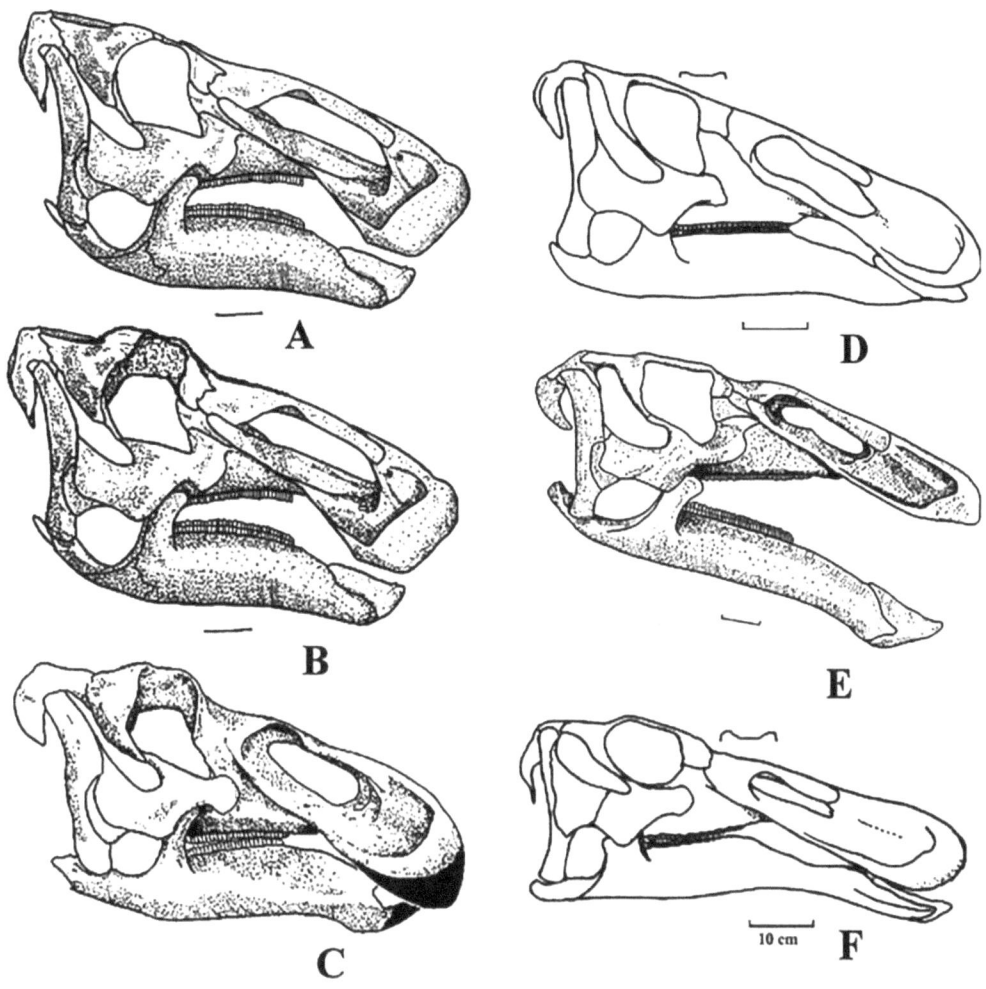

Figure 1) Skulls of Maastrichtian hadrosaurs drawn to the same height showing the differences in the skulls. A) Type of *Edmontosaurus regalis* (NMC 2288) drawn conservatively; B) Type of *Edmontosaurus regalis* (NMC 2288) corrected ridge over the eyes; C) *Edmontosaurus regalis* Field Museum specimen (CNHM P15003); D) *Anatosaurus* skull from the 2005 Tucson Rock Show; E) Uncrushed skull of *Anatotitan copei* (AMNH 5730); F) Skull of *Anatotitan* from the 2005 Tucson show.

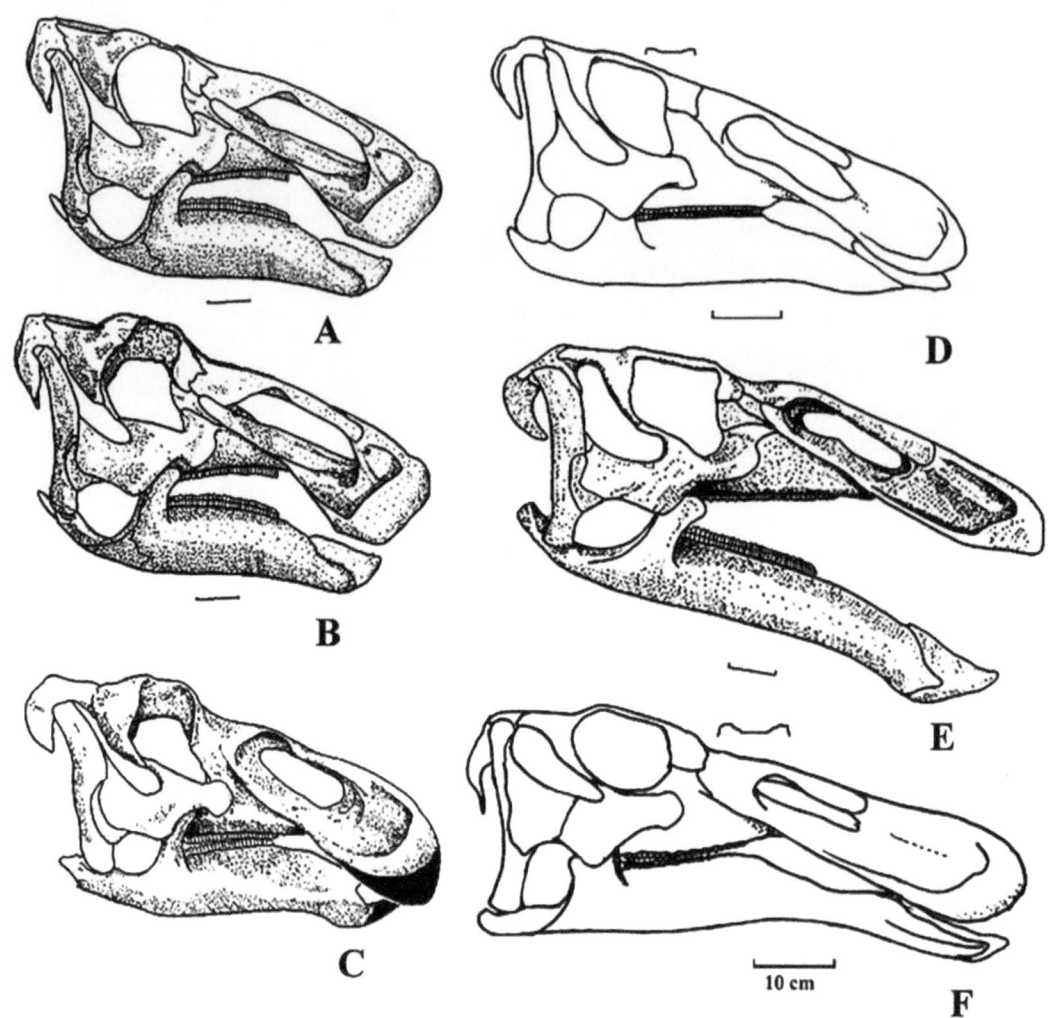

Figure 2) Skulls of Maastrichtian hadrosaurs drawn to the same height showing the differences in the skulls. A) Type of *Edmontosaurus regalis* (NMC 2288) drawn conservatively; B) Type of *Edmontosaurus regalis* (NMC 2288) corrected ridge over the eyes; C) *Edmontosaurus regalis* Field Museum specimen (CNHM P15003); D) *Anatosaurus* skull from the 2005 Tucson Rock Show; E) Uncrushed skull of *Anatotitan copei* (AMNH 5730); F) Skull of *Anatotitan* from the 2005 Tucson show.

The nearly complete skeleton of the mummified *Corythosaurus* at the AMNH (AMNH 5240) has a long tail, but not the whiplash seen in the above specimens. It is possible that lambeosaurs lacked the whiplash tail. A back half of a skeleton referred to *Corythosaurus casuaris* at the Smithsonian (USNM 15493) has a very long tail. It may be incorrectly referred to the lambeosaur. The tip of the pubis is hiding behind a leg, so you can't tell if it has a straight pubis or had a pubic boot. One of the best illustrated caudal series of a hadrosaur is *Tsintosaurus spinorhinus* (IVPP V725). It is an articulated series of 59 caudal vertebrae and it isn't a complete series. I've tried to extrapolate the complete length by using the angle of the decreasing height of the caudal centra. I came up with another 20 or so vertebrae making it series about 80 caudal vertebrae. This is more than what the Dinosauria II states which is 50-70 caudal vertebrae in hadrosaurs. Counting the vertebrae of *Corythosaurus* I come up with about 60-62 caudal vertebrae. USNM 15493 has just over 80 caudal vertebrae and this may be a good indication that it is not *Corythosaurus*. (Figure 4)

With a thick horse-like neck and large gullar pouch, a thick tail with the last 10 feet (or more) of a 'whiplash-like end and two thick sturdy hind legs and thin forelimbs, not to mention the bizarre shape of the crests, hadrosaurs were a more bizarre looking animal than is usually illustrated.

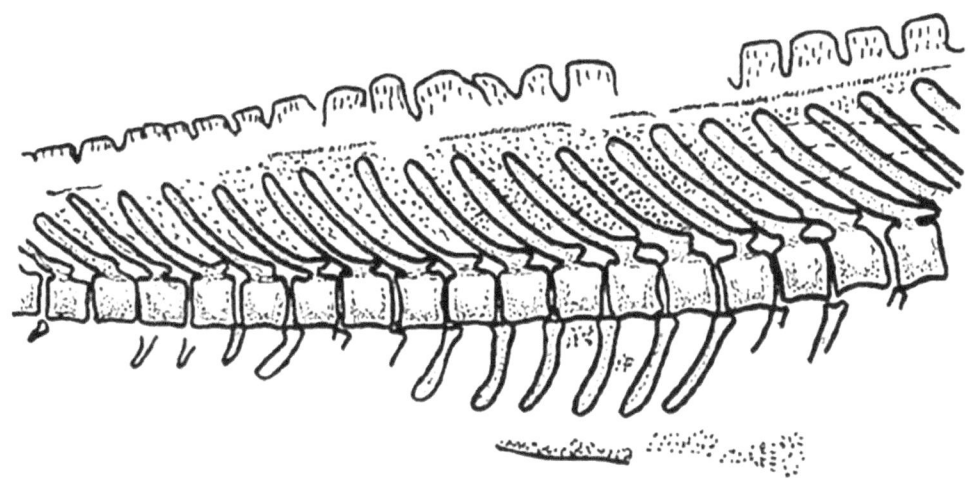

Figure 3) Tail of MOR V 007 showing the thickness of the tail on this half of the specimen.

Bibliography

Brown, B. B., 1913, The skeleton of *Saurolophus*, a crested duck-billed dinosaur from the Edmonton Cretaceous: Bulletin of the American Museum of Natural History, v. 32, p. 387-393.

Brown, B. B., 1916, *Corythosaurus casuarius*: skeleton, musculature and epidermis: Bulletin of the American Museum of Natural History, v. 35, p. 709-716.

Czerkas, S. A., 1993, Frills and goosenecks: Journal of Vertebrate Paleontology, v. 13, supplement to n. 3, Abstracts of Papers, Fifty-Third Annual Meeting, Society of Vertebrate Paleontology, New Mexico Museum of Natural History and Science, Albuquerque, New Mexico, October 13-16, p. 32A.

Horner, J. R., 1984, A "segmented" epidermal tail frill in a species of hadrosaurian dinosaur: Journal of Paleontology, v. 58, n. 1, p. 270-271.

Lambe, L. M., 1920, The Hadrosaur *Edmontosaurus* from the Upper Cretaceous of Alberta: Canadian Geological Survey Department of Mines, memoires 120, p. 1-79.

Lull, R. S., and Wright, N. E., 1942, Hadrosaurian Dinosaurs of North America: The Geological Society of America, Special Paper, n. 40, p. 1-242.

Manning, P. L., Morris, P. M., McMahon, A., Jones, E,. Gize, A., Macquaker, J. H. S., Wolff, G., Thompson, A., Marshall, J., Taylor, K. G., Lyson, T., Gaskell, S., Reamtong, O., Sellers, W. I., van Dongen, B. E., Buckley, M., and Wogelius, R. A., 2009, Mineralized soft-tissue structure and chemistry in a mummified hadrosaur from the Hell Creek Formation, North Dakota (USA): Proceedings of the Royal Society, Series B., v. 276, p. 3429-3437.

Rozhdestvensky, A. K., 1957, A duckbilled dinosaur, a saurolph from the Upper Cretaceous of Mongolia: Vertebrata PalAsiatic, v. 1, n. 2, p. 129-149.

Young, C.-C., 1958, The Dinosaurian Remains of Laiyang, Shantung: Palaeontologia Sincia, Whole Number 142, new series C, n. 16, p. 1-138.

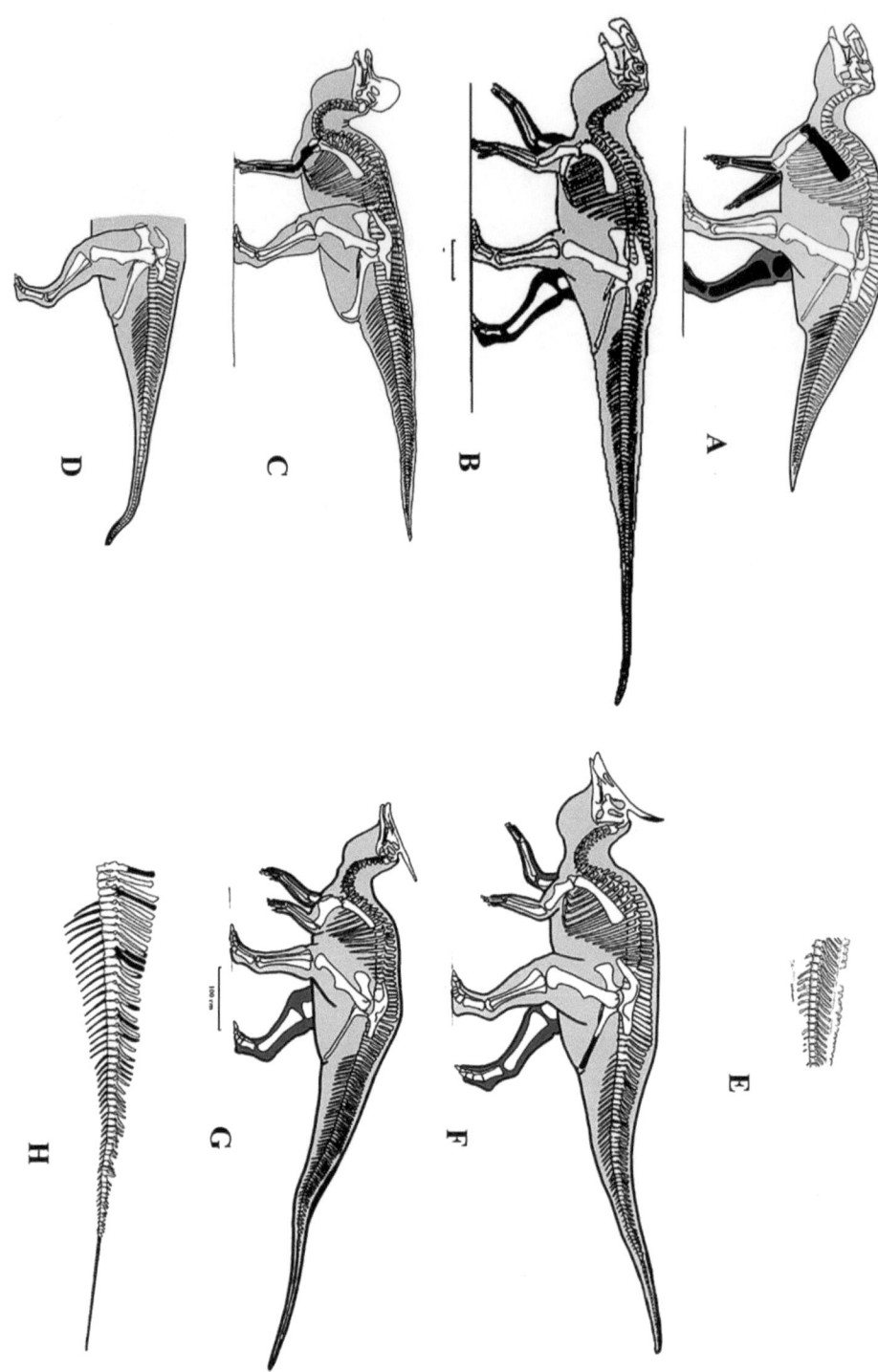

Figure 4) Skeletons of hadrosaurs showing the length of the tail; A) *Edmontosaurus regalis* (NMC 2288); B) Skeleton of *Anatosaurus* showing the new interpretation of the thickness of the neck and tail and the whiplash like tail; C) Skeleton of *Corythosaurus casuaris* (AMNH 5240); D) referred tail section of *Corythosaurus casuaris* (USNM 15493); E) MOR V 007 tail section; F) *Saurolophus osborni* (AMNH 5220), G) *Saurolophus anustirostris* (PIN 551-8?), H) *Tsintosaurus spinorhinus* (IVPP V725).

Ford, T. L., 2008, How to Draw Dinosaurs, Swimming with dinosaurs, part 1, Prehistoric Times, n. 86, p. 18-19.

Chapter 14

Swimming with dinosaurs, part 1

Aquatic dinosaurs were the norm in the early 20th century. Lumbering sauropods were believed to only have been able to sustain their bulk if they lived in lakes. Hadrosaurs with their bizarre hollow crests were believed to have lived in lakes and rivers with their bizarre hollow crests storing air as they swam and fed on aquatic vegetation. These images were not only iconic for the public but were accepted by paleontologists. It wasn't until the late 1960's and early 1970's that this lifestyle was challenged with the 'Dinosaur renaissance'. Not only did the reinterpretation take dinosaurs out of the swamps, rivers and lakes but also showed they weren't the slow lumbering, sluggish reptiles they were depicted to be. But the removal of dinosaurs from freshwater habitats leaves that ecological niche vacant of medium to large sized animals. I gave a talk at a Mesa Southwest Museum Symposium in 2002 about the possibility of some ankylosaurs having a lifestyle similar to Hippopotamus.

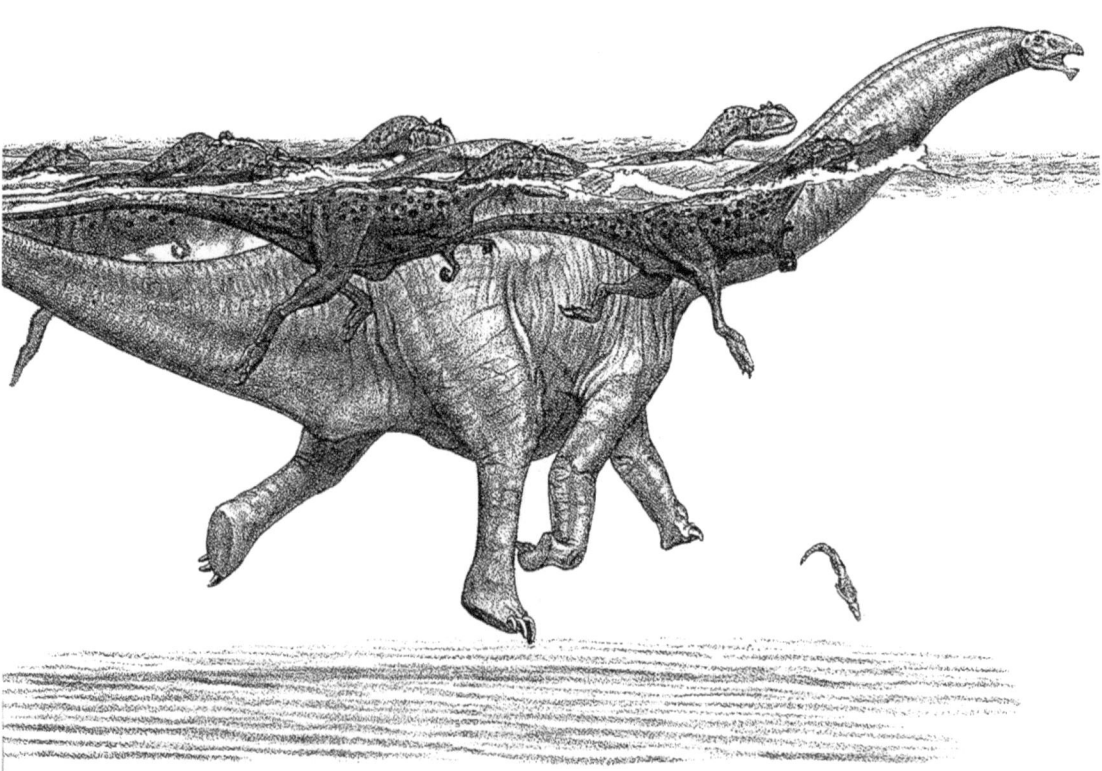

Figure 1) Greg Paul's illustration of a Sauropod being chased by Allosaurs (with permission from Greg Paul).

In this series I will be looking at the growing evidence that some groups of dinosaurs may have been semi- to aquatic animals, by looking at both the ichnological and fossil evidence. I'm co-authoring a paper with Larry Martin about the possibility of some if not all psittacosaurs being semi-, or aquatic animals. This will be published in the forthcoming Ceratopian volume by Indian University Press. I'm hoping I can time the next part to come out just after that book. Also, a new paper by Tereschenko is about the possibility that some protoceratopians were semi-, or aquatic animals and I'll be commenting on that at in the same article.

For some reason paleontologist are reluctant to put dinosaurs into niches other than purely terrestrial. For example, climbing or arboreal dinosaurs was unheard of a decade ago, but now more and more are excepting the strong possibility of climbing by some dromaeosaurs. Burrowing ornithopods was unheard of a decade ago (though Bakker in 1990 did argue for *Drinker* being a burrower), and now burrowing ornithopods are known (*Oryctodromeus*, as I talked about in PT 82). There may be other known ornithopods that were also burrowers, though not recognized as so, but more research needs to be done to confirm this. This also goes for aquatic or semi aquatic dinosaurs. A semi aquatic dinosaur seems impossible if not improbable for many paleontologists.

The first group of dinosaurs I'll be talking about are theropods. When Bakker (1971) revolutionized the habitat of sauropods from amphibious to terrestrial animals also stated that theropods, such as *Allosaurus* and *Megalosaurus,* were better swimmers. They probably swam in a similar manner as large ground-living bird; emus, rheas, and cassowaries. These large birds are good swimmers and use their large powerful legs and long toes to swim.

Greg Paul, independently, came to the conclusion that theropods were better swimmers than sauropods and illustrate a swimming sauropod being chased by a group of allosaurs (Figure 1).

Walter Coombs was the first to describe swimming theropod tracks from the Lower Jurassic of Rocky Hill, Connecticut. He attributed the tracks to *Eubrontes* sp. He believed the tracks were from a swimming theropod that had their feet perpendicular to the substrate, dug its claws straight down into the substrate and then kicked off. He has an illustration showing this. This find though, has recently been shown to be just poorly preserved prints and not from a swimming theropod (Figure 2).

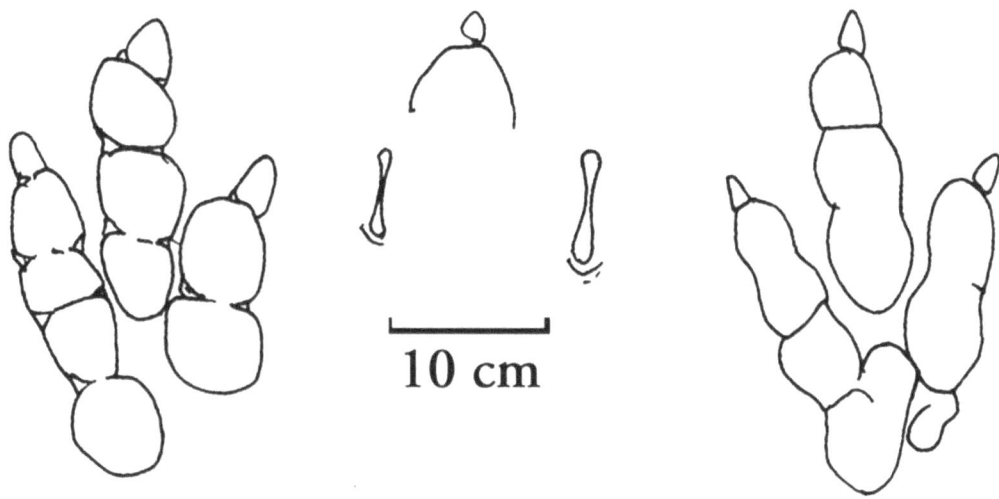

Figure 2) Coombs swimming theropod tracks, *Eubrontes* sp.

But it turns out Coombs was right, (just that he was wrong about the Connecticut tracks). Several other true swimming traces have been found from around the world and from the Late Triassic to the Early Cretaceous. Swimming traces are just the claw marks of a theropod as it scraps its claws on the substrate. At the St. George Dinosaur tracksite in Utah are extremely good and important tracks and trackways of several kinds of animals. The trackways come from several different layers of the Early Jurassic Moenave Formation. These theropod trackways go from close to shore with well established three toed tracks to just claw marks indicating a swimming animal. The swimming traces are sets of three parallel scrape marks that taper at each end, with the longer middle toe leaving a longer and deeper scrap mark (Milner & Kirkland, 2007). The Johnson Farm sandstone bed has the most abundant swim traces (called *Characichnos*). They are believed to be from a grallator like theropod that were stepping from near shore and into deeper water where it was completely swept off its feet (Figure 3). The authors wonder why there are so many swimming traces? The answer may be in the fish fauna. There are numerous fish remains at this site; hypodont sharks, the lungfish *Ceratodus*, a coelacanth similar to *Chinlea* and semionotid fish as big as 4 feet in length. Thus, they may have been trying to catch fish. They also believe teeth from the area are similar to *Dilophosaurus* and these long thing teeth are similar to fish eating reptiles. This early Jurassic site may be the earliest evidence for fish eating theropods. They also believe that *Dilophosaurus* had some similarities to spinosaurids (something that I've believed for more than a decade). The nares are placed further back on the skull than other theropods, the premaxilla is expanded, with long teeth, and they had long strong arms. Milner and Kirkland are quick to say that they aren't saying that those teeth are from a spinosaurid, just that they have similarities to them and are from a totally different theropod (Figure 5).

Figure 3) Johnson Farm trackway illustration. Two theropods, one wading and the other swimming (modified from Milner & Kirkland, 2007).

Recently a theropod trackway from the Early Cretaceous of Spain has been found. The trackway has 12 consecutive tracks of a large swimming theropod. It is believed that the animal was swimming in 3.2 meters of water. The theropod would have had a hip height of about 3.5 meters (about 11 feet). The trackway indicates a large sized theropod was swimming at an angle against the current. Could this trackway have been from a baryonyxid? The current was swift, and it appears the theropod had a hard time swimming (Figure 4).

Baryonyx is believed to have caught fish (which really isn't out of the question because of acid etched scales and teeth of a *Lepidotes* fish was found in its stomach area along with acid etched *Iguanodon* bones). Many believe that they used their large thumb claw to catch fish in a similar manner as modern bears, standing along the shore or slightly wading into the water and when a fish swam by the snatched it with its claw. But the problem with that is the skull and neck would have been a better tool for catching fish. It has a long low skull with large round front teeth, and a highly mobile neck that would have made catching fish all the more easily than it would have it just used its claws. Recent studies comparing the skull of *Baryonyx* and for that matter spinosaurids in general, show that they have more in common with the gharial and other long slender snouted crocodiles that with broader skulled crocodiles or theropods. Because the skulls are so long, and have long premaxillary and dentary teeth, it is believed they caught their prey with the tip of their snout, presumably fish. Another possibility that the authors raise is scavenging. The more recessed naries and long snout would make it ideal for digging into a carcass, but the skull could not withstand the torsion that the animal would have had when shaking the head side to side to rip out chunks of meat (Figure 5).

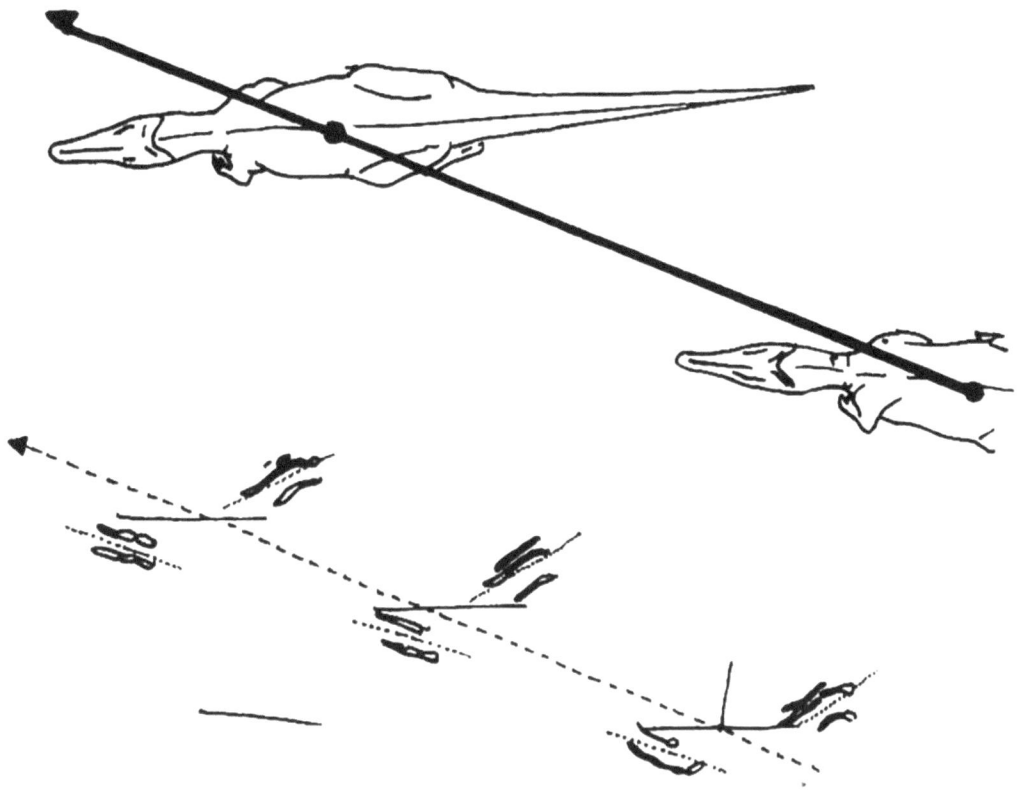

Figure 4) Spanish swimming theropod. The solid and dotted line the direction the theropod was swimming. The left hand side is the tracks, the right the theropod swimming at an angle against the current (From Ezquerra et al, 2007).

The problem with *Baryonxy*/spinosaurs catching fish by crouching on the banks is that the size of the fish that they are believed to have caught were medium to large fish, which wouldn't have been close to the shore and would have lived in deeper water. Spinosaurids are interesting theropods; with higher than normally placed orbits, nares that are placed further back on the skull, and the occipital condyles was held horizontally which means the head was held at the same level as the neck. The neck didn't have the typical theropod 'S' shape and was held horizontally to the body which means the head, neck, body and tail was held in a straight line. Though the neck was held horizontal it was very mobile which may have helped in catching a swimming animal. The fore limbs were very strong with a large 'thumb' claw that was either used for gaffing or ripping into a carcass. The morphology of the teeth, skull and neck indicate they couldn't rip into prey like other theropods and used its strong forelimbs with the large thumb claw to do so. It may have also been possible for spinosaurids to use their powerfull forelimbs in swimming. The tails had tall neural spines, were laterally compressed and may have been used for propulsion. Because the skull, neck, body and tail were held horizontally may be a hindrance in a completely terrestrial animal, but not a swimming one. Perhaps spinosaurids were like large crocodilians and lived more in water than land to feed than like typical terrestrial theropods? (Figure 6).

Swimming in large lakes/rivers with their eyes barely exposed, the majority of their body under water, using their powerful forelimbs and tail to swim and using their highly mobile neck and long skulls may have been ideal for them to catch large fish. Nonetheless, something that works against a swimming habit it the skeleton has a typical 'terrestrial' theropod morphology with long hind legs and long dorsally flattened ribs.

Baryonyx was the first good specimen of a spinosaurid with skull material. Sereno et al, (1998) described *Suchomimus* from Africa which had a little more skull elements. It is interesting that the lower

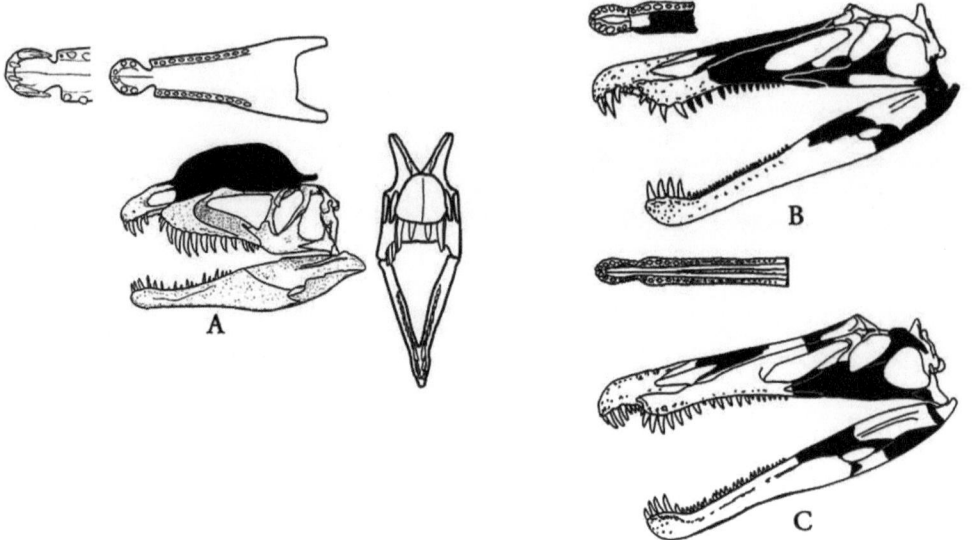

Figure 5) Skulls of *Dilophosaurus* and Baryonyxids; A) *Dilophosaurus*, in ventral, side, and front view (The ventral and front view is the first time they have been illustrated, anywhere); B) *Baryonyx walkeri*; C) *Baryonyx* (*Suchomimus*) *tenerensis*. I have redrawn both these skulls by using the new spinosaur skulls.

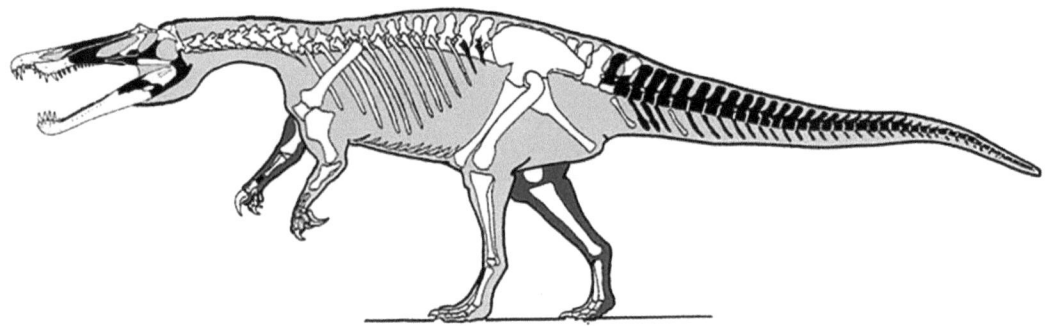

Figure 6) New skeleton illustration of *Baryonyx walkeri*.

Figure 7) Illustration of two swimming *Baryonyx walkeri*.

jaw of both has the same amount of teeth preserved as well as being nearly the same size. To me this means the skulls should be about the same size and that *Baryonyx* didn't have a shorter skull than *Suchomimus* (Figure 5). Also, I believe that Milner (that is the British A. C. Milner) assertion that *Suchomimus* is actually a *Baryonyx* is correct.

More and more spinosaur specimens are being found and it is believed that many if not all spinosaurs were fish eaters or that fish were just part of their menu. I believe in order to catch fish they'd have had to have gone into a river, pond, lake or stream to catch them and not just stand on the shore or wade into shallow water to catch fish. There may be other theropods that also may have been fish eaters and either have not been found or not recognized for a piscivorous life.

Bibliography

Bakker, R. T., 1971, Ecology of the Brontosaurs: Nature, v. 229, p. 172-174.

Charig, A. J., and Milner, A. C., 1997, *Baryonyx walkeri*, a fish-eating dinosaur from the Wealden of Surrey: Bulletin of The Natural History Museum, Geology Series, v. 53, n. 1, p. 11-70.

Coombs, W. P. jr., 1980, Swimming Ability of Carnivorous Dinosaurs: Science, v. 207, p. 1198-1200.

Ezquerra, R., Doublet, S., Costeur, L., Galton, P. M., and Perez-Lorente, F., 2007, Were non-avian theropod dinosaurs able to swim? Supportive evidence from an Early Cretaceous trackway, Cameros Basin (La Rioja, Spain): Geology, v. 35, n. 6, p. 507-510.

Milner, A. C., 2004, *Baryonyx*, a fish-eating dinosaur (Theropoda: Spionsauridae) from southern England and the palaeobiology and palaeogeography of the spinosaurids: In: British Dinosaurs, A Palaeontological Association Review Seminar, Co-hosted by Dinosaur Isle Museum and the University of Portsmouth, School of Earth and Environmental Sciences, Programme and abstracts, unnumbered.

Milner, A. R. C., and Kirkland, J. I., 2007, The case for fishing dinosaurs at the St. George dinosaur discovery site at Johnson Farm: Utah Geological Survey, Survey Notes, v. 39, n. 3, p. 1-3.

Tereschenko, V. S., 2007, Key to protoceratopoid vertebrae (Ceratopsia, Dinosauria) from Mongolia: Palaeontological Journal, v. 41, n. 2, p. p. 175-188.

Welles, S. P., 1984, *Dilophosaurus wetherilli* (Dinosauria, Theropoda) osteology and comparisons: Palaeontographica Abt. A 185, lfg. 4-6, p. 85-180.

Ford, T. L., 2008, How to Draw Dinosaurs, Swimming with dinosaurs, part 2, Prehistoric Times, n. 87, p. 18-19.

Chapter 15

Swimming with dinosaurs, part 2

Well, my co-authored article with Larry Martin about the possibility of swimming *Psittacosaurus* is as of this writing, in press (I think), so hopefully next time I'll be able to write about that intriguing but not new theory. This time I'm talking about Sauropods!

I find it interesting that in 1878 Marsh believed that *Titanosaurus montanus* was a terrestrial animal (he didn't know at the time the name *Titanosaurus* was already used the by Lydekker that same year. When he found out about the preoccupied name latter that same year Marsh replaced his *Titanosaurus* with *Atlantosaurus*) …'One of these monsters (*Titanosaurus montanus*), from Colorado, is by far the largest land-animal yet discovered; its dimensions being greater than was supposed possible in an animal that lived and moved upon the land. It was some fifty or sixty feet in length, and, when erect, at least thirty feet in height! (I wonder if he was thinking it could stand on its hind legs?) It doubtless fed upon the foliage of the mountain forests, portions of which are preserved with its remains…but in 1883 he believed *Brontosaurus* was aquatic…In habits, *Brontosaurus* was more or less amphibious, and its food was probably aquatic plants or other succulent vegetation. The remains are usually found in localities where the animals had eventually become mired…in 1884…The position of the external nares indicates an aquatic life (he was talking about *Diplodocus*)… But their shear bulk helped in the interoperation that sauropods had an aquatic lifestyle. But all of this was challenged during the dinosaur renaissance of the early 1970's and they were taken out of the water and placed on land becoming a completely terrestrial animal. And this has been the ongoing theory by paleontologist and artist since then. But were they truly just terrestrial dinosaurs?

The beautiful painting by Burian with the *Brachiosaurus* standing in 40 feet of water was then obsolete. Research showed that just the pressure of the water itself on the rib cage would make it impossible for them to breathe and like theropods and birds, some of their vertebrae were pneumatic and had air sacs. This pneumaticity has been known for as along as sauropods. Marsh (1877) believed they were a weight reduction, while his rival Cope (1878) thought that the vertebrae would have been used as 'floats'. Pneumatic vertebrae would not only lighten the animal as a whole but make it impossible for them to sink. Even though sauropods are the largest terrestrial animal, they would have bobbed in the water.

There are things to look at to help determine if they lived on land or water or something that combines the two. Two of which are ichnology and paleoenvironments.

What got me thinking about this again was *Paralititan*. A few years ago, a talk was given at the SVP in which the author described the whole northern end of Africa as a kind of mangrove swamp. New research on sedimentology and paleo-environments also supports this. The researchers looked into whether or not the wet substrate could in fact hold its bulk and the conclusion was yes. Even though the area was a mangrove-like swamp, the substrate was sandy not muddy, and the sandy substrate could easily hold the bulk of the sauropod. This would mean *Paralititan* would have been walking and living in a swamp/mangrove-like environment. Just image a *Paralititan* wondering through the swamp, trees and waterways feeding and avoiding *Spinosaurus*, *Carcharodontosaurus*, large crocodiles, and large coelacanths. How this relates to the Morrison or other areas I don't know. The Morrison was as far as I know, a more arid environment with flood plains that had at times been inundated with water from storms (?) that helped move carcasses from the plain into rivers. It is possible that not all sauropods lived in the same environment.

It also turns out that sauropods like to lay eggs and made nests in distal alluvial or coastal plains, or even tidal flats. This is true for nests that have been found in India, South America and Europe. As far as I know no sauropod nests are known from North America. Why sauropods liked to make nests in these moist environments I don't know. Perhaps it was easier to dig nests. Hadrosaurs liked a more arid environment to lay their eggs and unlike hadrosaur nests no foliage were found in the sauropod nests.

There are several sauropod trackways that are made up with just the manus prints and no pes prints, and conversely there are trackways with just the pes prints. This has been interpreted as being underprints by some, but this is more than likely true manus prints and are believed to be from swimming sauropods. Just manus tracks are known from the Middle Jurassic of Morocoo (Figure 1 from Ishigaki,

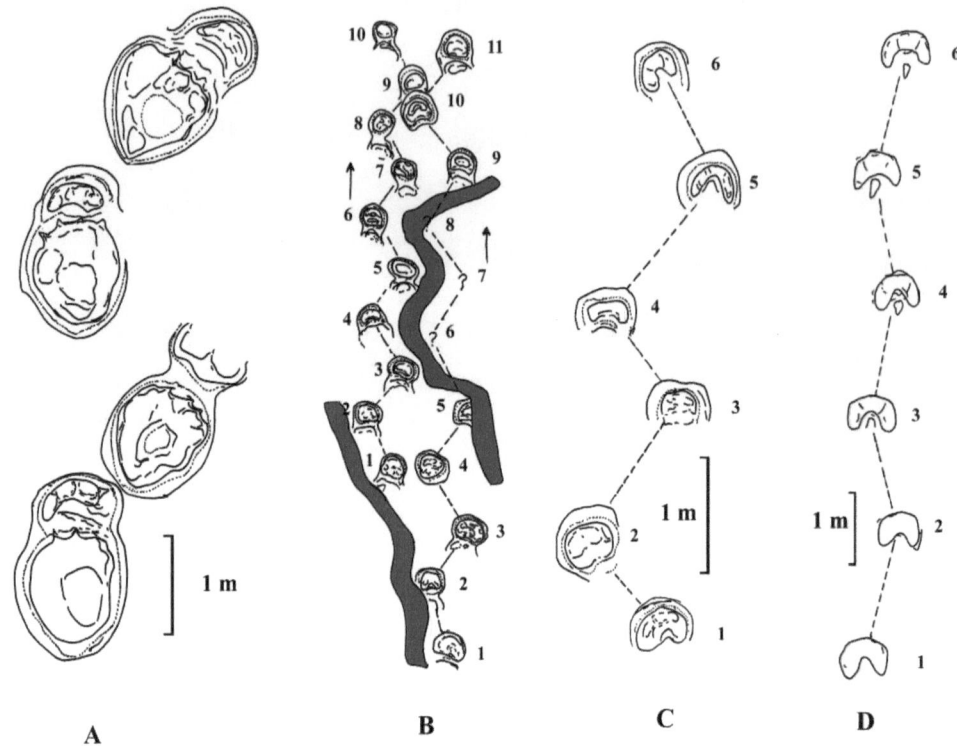

Figure 1) Sauropod trackways with manus and pes prints and just manus prints all from Ishigaki, 1989; A) *Breviparopus taghbaloutensis*; B-C) Trackways from Morocco; B) Trackways A and B; C) Trackway C; D) Trackway D.

1989) and the Late Cretaceous of North Korea. Just pes prints are also known from Mexico and was believed to indicate bipedal sauropods, but there is a possibility that they are also from swimming sauropods.

If these tracks are from swimming sauropods was if physically possible for them to swim with either the front or hind legs? Did different families of sauropods swim differently? Donald M. Henderson addressed this a few years ago (2004). He studied the pneumatic vertebrae, length of limbs, position of the vertebral column, center of balance and center of mass and came up with how sauropods could have swam. He calculated how much bulk it would have had and where its buoyancy point was when it was in the water as well as plotting the center of balance, center of mass vs. buoyancy and came up with two different scenarios for swimming sauropods.

Sauropods had a narrow chest and if they could not touch the bottom of a river/lake/pond etc, they would have flipped onto their sides (as per Henderson), but their long necks would have kept them from completely capsizing. But if they could keep at least two feet on the substrate, they could keep upright. But since there are no living sauropods it is impossible to completely state that sauropods couldn't have swam completely free from the substrate. They may have used their muscles, limbs, and body movements to help keep them upright when they swam.

The long front legged brachiosaurs/camarasaurs hind feet would have lifted off the substrate before the front would have. The center of balance in brachiosaurs/camarasaurs was posteriorly positioned with the center of mass being in the middle of the body in *Brachiosaurus* (because they had longer bodies than other sauropods) and more toward the hind legs in *Camarasaurus*. This may be the sauropod families that could have made just the manus trackways.

Sauropods with shorter front legs and longer hind legs would have had the front legs raised off the substrate first. Diplodocids have the center of mass and balance nearly at the front of the hind legs. The trackways with just pes prints could be from a swimming diplodocid and not a bipedal walking sauropod (like I illustrated before in PT). I don't know if those trackways were looked at to see if they could tell how deep the water was because there had to have been some sort of water for the track to have been formed.

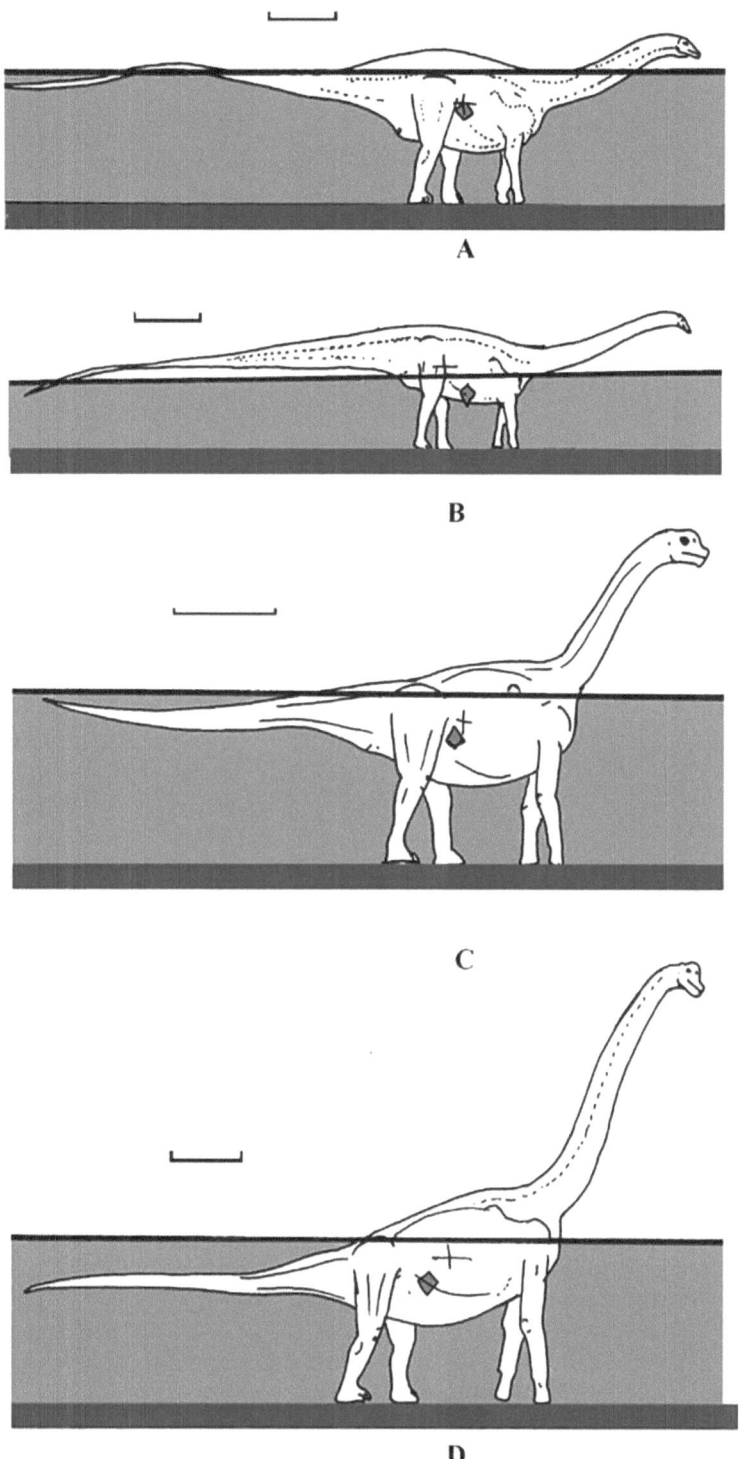

Figure 2) Sauropods standing in water just under their equilibrium depths all after Henderson, 2004. Open diamond marks the center of buoyancy, black plus symbol marks the centre of mass. A) *Apatosaurus* critical depth 3.7m; B) *Diplodocus* at 2.4 m; C) *Camarasaurus* at 3.2 m; D) *Brachiosaurus* at 4.3m.

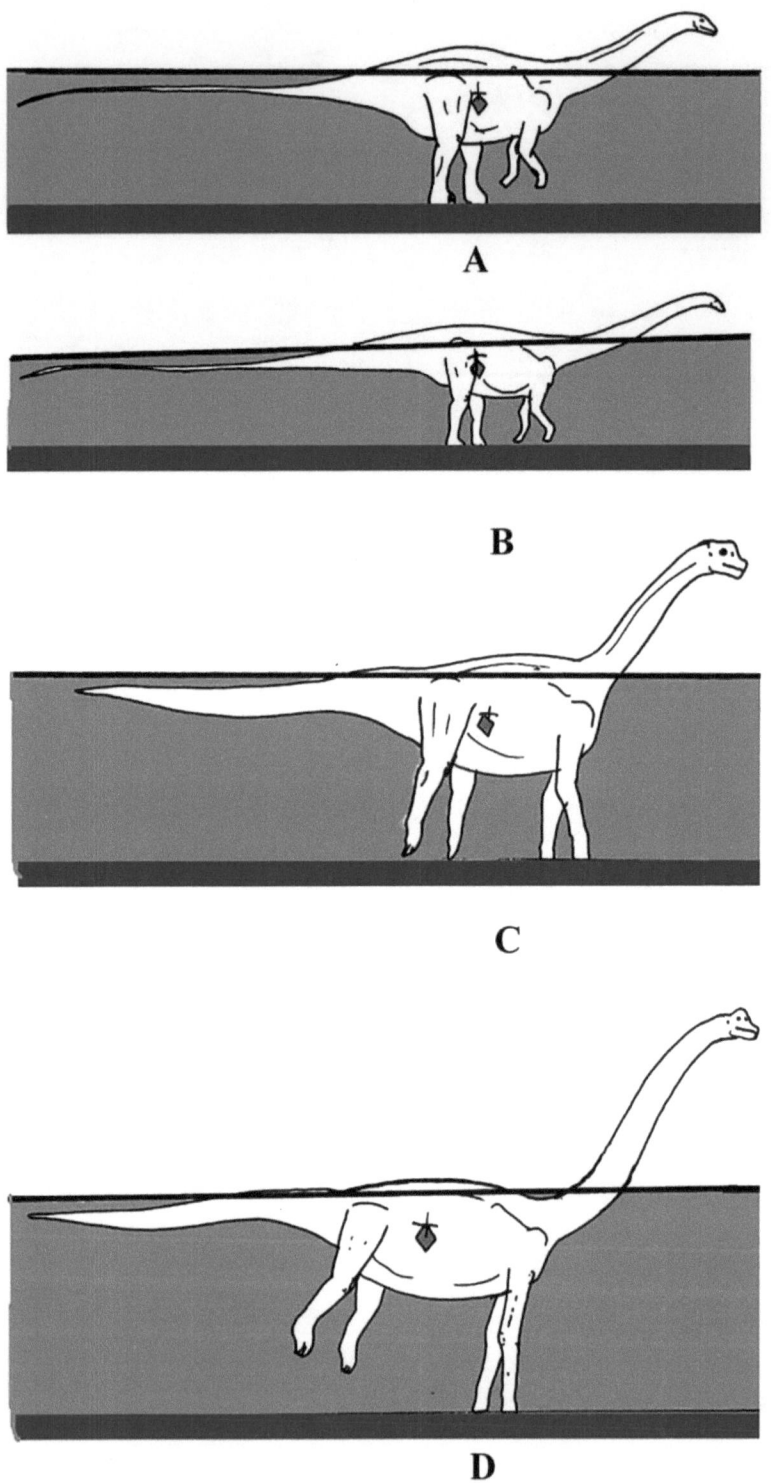

Figure 3) Sauropods swimming with a pair of limbs on the substrate all after Henderson, 2004. Open diamond marks the center of buoyancy, black plus symbol marks the centre of mass. A) *Apatosaurus*; B) *Diplodocus*; C) *Camarasaurus*; D) *Brachiosaurus*.

Ichnology, nests, sedimentology indicates sauropods frequented moist/watery areas. Did they prefer that kind of environment? Did some live a completely terrestrial lifestyle like those sauropods found in the Late Cretaceous Mongolian Nemegt Formation? Did others live along shorelines (whether freshwater or salt) as European, South American and Indian trackways and nests indicate? If some sauropods did prefer living along shorelines, then why hasn't there been any footprints/trackways found along the Middle to Late Cretaceous Interior seaway in North America? Perhaps the shoreline hasn't been found yet.

Does this mean that sauropods were aquatic? Or were they just crossing a water source to basically get to the other side? More research needs to be done, but completely terrestrial sauropods needs to be looked into again and the possibility of swimming sauropods, like that in which I grew up with, needs to be readdressed. The great Zallenger painting at Yale (or the one that use to be at Yale?) of a *Brontosaurus* standing in a river (like the copy I have hanging on my wall) may not be so far fetched...

Bibliography

Cope, E. D., 1878, A new opisthocoelian dinosaur: American Naturalist, v. 12, p. 406.

Henderson, D. M., 2004, Tipsy punters: sauropod dinosaur pneumaticity, buoyancy and aquatic habits: Proceedings of the Royal Society of London, Series B, supplement, n. 271, p. S180-S183.

Henderson, D. M., 2004, Sauropod dinosaur—the colossal corks of the Mesozoic: Alberta Palaeontological Society, Eighth Annual Symposium, "Fossils Unwrapped", Abstracts, p. 20-21.

Ishigaki, S., and Fujiskai, T., 1989, Three Dimensional Representation of *Eubrontes* by the Method of Moire Topography: In: Dinosaur Tracks and Traces, edited by Gillette, D. D., and Lockley, M. G., Cambridge University Press, p. 421-425.

Lacovara, K. J., Smith, J. B., Smith, J. R., Lamanna, J. B., Johnson, K. R., Lyon, M. A., Dodson, P., and Nichols, D. J., 2001, Coastal environments of Cretaceous dinosaurs: examples from North Africa and western North America: Journal of Vertebrate Paleontology, v. 21, supplement to n. 3, abstracts of papers, Sixty-first annual meeting, Society of Vertebrate Paleontology, Museum of the Rockies, Montana State University, Bozeman, Montana, October, 3-6, p. 70a.

Marsh, O. C., 1877, Notice of a new gigantic dinosaur: American Journal of Science, 3rd series, v. 14, p. 87-88.

Marsh, O. C., 1878, Principal Characters of American Jurassic Dinosaurs, Part I: American Journal of Science, 3rd series, v. 16, p. 411-416.

Marsh, O. C., 1883, Principal Characters of American Jurassic Dinosaurs. Part VI. Restoration of *Brontosaurus*: American Journal of Science, 3rd series, v. 27, p. 81-85.

Ford, T. L., 2008, How to Draw Dinosaurs, Were Stegosaurs bipedal? Prehistoric Times, n. 88, p. 18-19.

Chapter 16

Were Stegosaurs bipedal?

This isn't a new idea. In fact, it the idea/theory goes as far back as late 1800's, and possibly further than that. Frank Bond in 1899, under the direction of Charles Knight, illustrated a bipedal stegosaur (Figure 1a), and recently, this topic of bipedal stegosaurs has been renewed by Gierlinsky and Sabath (2008). They reported on some stegosaur tracks, both definitive tracks and tracks that were misidentified as being from a stegosaur. Stegosaur feet and tracks are unique. The pes has the typical ornithischian 3 toes and the manus has 5 toes. The toes are short and stubby (*Stegosaurus* manus phalangeal formula is 2-2-2-2-1 including phalange and ungual and the pes has 0-2-2-2-0, Figure 2) and therefore stegosaur tracks are easily identified. Lockley and Hunt (1998) named stegosaur tracks (*Stegopodus czerkasi*) in honor of Stephen Czerkas for his work on *Stegosaurus*. The first tracks were from Salt Valley near Moab Utah and more tracks were described from the Cleveland-Lloyd Quarry by Gierlinsky and Sabath. Both set of tracks are from the Morrison Formation. They also describe additional stegosaur tracks from the Tenes Cliffs in Asturias, Spain (Tenes Formation, Late Jurassic). The tracks from Spain are the only unquestionable bipedal Stegosaur tracks.

In an article by Sargent (2001) he talks about an illustration by C. H. Murry Chapman who in 1924 illustrated a *Stegosaurus* in a bipedal pose (Figure 1b). But as I said Bond illustrated a bipedal stegosaur back in 1899 (it is standing on its hind legs and but is leaning on a tree with its front legs). If you look closely (unless the illustration is really small), *Stegosaurus* is illustrated with the plates lying flat on the back and the spikes intermingled between them. The reason for this is that when Marsh first described *Stegosaurus* (*S. armatus*), he thought the plates were like those of a turtle and were interlocking and layered on the back and sides. Its name does men roofed lizard but roofed as in a roof over a house. Stego ((s)teg), in Greek means cover as in 'cover with a roof'. It wasn't until *Stegosaurus stenops* was discovered that he realized the plates stood up in double rows down the middle of the back.

Stegosaurs have long hind legs and short front legs. There limb proportions are different in not only genera, but also in different species. *Stegosaurus ungulatus* has the longest hind legs to front legs while *Stegosaurus stenops* hind legs are shorter, but still longer then the front. A recent article by Maidment, Norman, Barret and Upchurch (2008) combined all the species of *Stegosaurus* in to the type species *S. armatus* and four other genera (*Hesperosaurus, Diracodon, Hypsirophus and Wuerhosaurus*). I have no problem with *Stegosaurus stenops* being sunk into *Stegosaurus armatus* (after all they're from the same quarry) but disagree with the placement of *Stegosaurus ungulatus* and other four genera (though *Diracodon* is more than likely a species of *Stegosaurus*). Maidment et al do say they share some characteristics, but there are also differences. If *Wuerhosaurus* was a *Stegosaurus*, then that would push the chronological age of *Stegosaurus* to 10mya, which I believe wasn't the case.
As per Maidment et al, (2008)

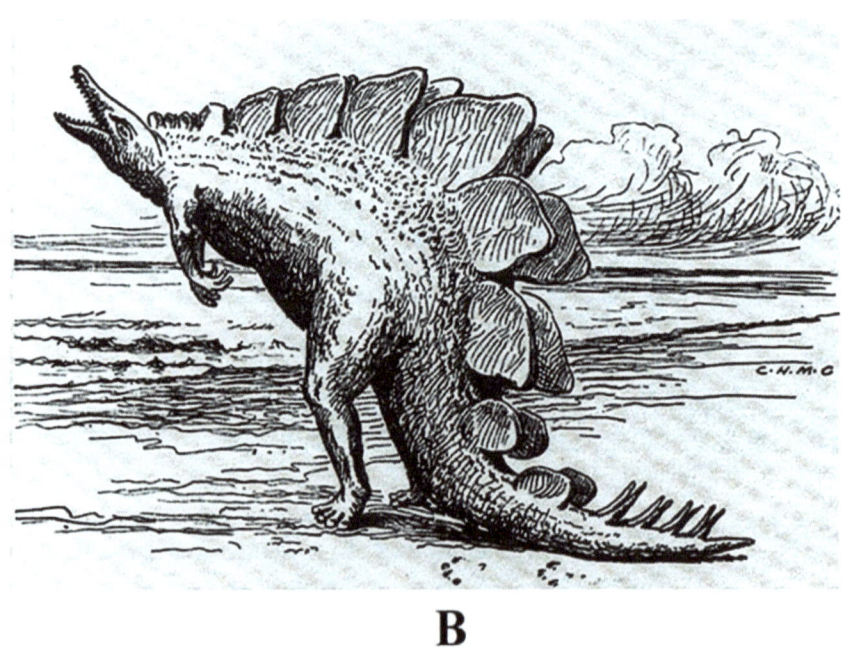

Figure 1) Early illustrations of bipedal stegosaurus A) Frank Bond, 1899; B) C. H. Murry Chapman, 1924.

Genus: *Stegosaurus* MARSH, 1877
= *Hypsirophus* COPE, 1878
= *Diracodon* MARSH, 1881
= *Wuerhosaurus* DONG, 1973
S. armatus MARSH 1877 (Type)
= *Stegosaurus stenops* MARSH, 1887
= *Diracodon stenops* (MARSH, 1887) BAKKER, 1986
= *Hypsirophus discurus* COPE, 1878
= *Stegosaurus discurus* (COPE, 1878)
= *Hypsirhophus seeleyanus* COPE 1879 (*nomen nudum*)
= *Stegosaurus seeleyanus* (COPE 1879) (*nomen nudum*)
= *Stegosaurus ungulatus* MARSH 1879
= *Diracodon laticeps* MARSH, 1881
= *Stegosaurus laticeps* (MARSH, 1881)
= *Stegosaurus duplex* MARSH 1887
= *Stegosaurus sulcatus* MARSH 1887
S. homheni (DONG, 1973) MAIDMENT, NORMAN, BARRETT & UPCHURCH, 2008
= *Wuerhosaurus homheni* DONG, 1973

Genus: *Stegosaurus* non MARSH, 1877
S? affinis MARSH 1881 (*nomen nudum*)
S. longispinus GILMORE, 1914

Genus: *Wuerhosaurus* non DONG, 1973
W. ordosensis DONG, 1993

I believe the following

Genus: *Alcovasaurus* GALTON, & CARPENTER, 2016
= *Natronasaurus* ULANSKY, 2014a (*nomen nudum*)
A. longispinus (GILMORE, 1914) GALTON, & CARPENTER, 1914 (Type)
= *Stegosaurus longispinus* GILMORE, 1914
= *Natronasaurus longispinus* (GILMORE, 1914) ULANSKY, 2014a (*nomen nudum*)

Genus: *Diracodon* MARSH, 1881 (*nomen dubium*)
D. laticeps MARSH, 1881 (Type)
= *Stegosaurus laticeps* (MARSH, 1881)

Genus: *Hypsirophus* COPE, 1878
H. discurus (COPE, 1878)
= *Hypsirophus discurus* COPE, 1878
= *Stegosaurus discurus* (COPE, 1878)
= *Hypsirhophus seeleyanus* COPE 1879 (*nomen nudum*)
= *Stegosaurus seeleyanus* (COPE 1879) (*nomen nudum*)

Genus: *Stegosaurus* MARSH, 1877
S. armatus MARSH 1877 (Type)
= *Stegosaurus stenops* MARSH, 1887
= *Diracodon stenops* (MARSH, 1887) BAKKER, 1986
S. ungulatus MARSH 1879
= *Stegosaurus duplex* MARSH 1887
S. sulcatus MARSH 1887

Genus: *Wuerhosaurus* DONG, 1973
= *Wuherosaurus* SERENO, 1986 (*sic*)
W. homheni DONG, 1973 (Type)
= *Stegosaurus homheni* (DONG, 1973) MAIDMENT, NORMAN, BARRETT & UPCHURCH, 2008
W. ordosensis DONG, 1993

Genus: *Stegosaurus* non MARSH, 1877
S? affinis MARSH 1881 (*nomen nudum*)

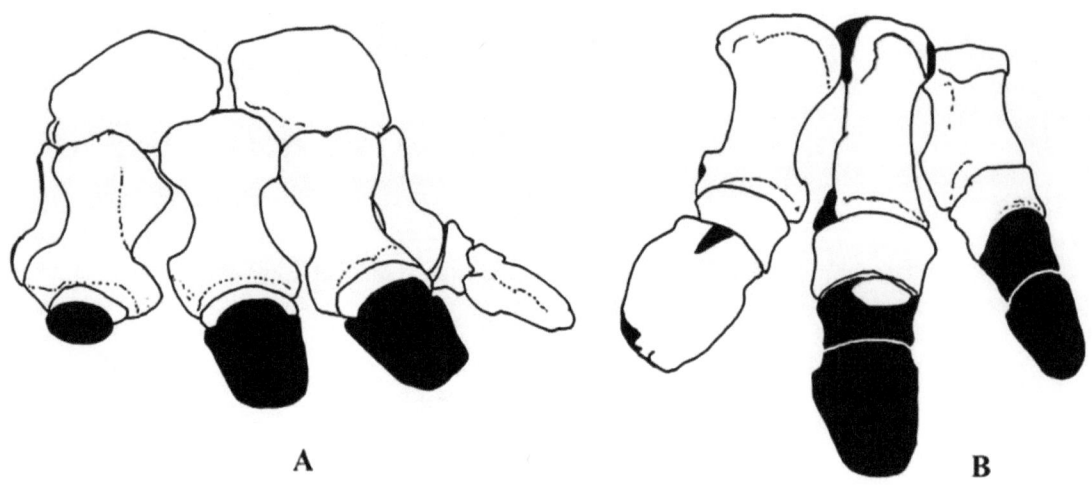

Figure 2) Manus (A, *Stegosaurus sulcatus*, USNM 4397) and Pes (B, *Stegosaurus* sp USNM 4280) skeleton of *Stegosaurus*.

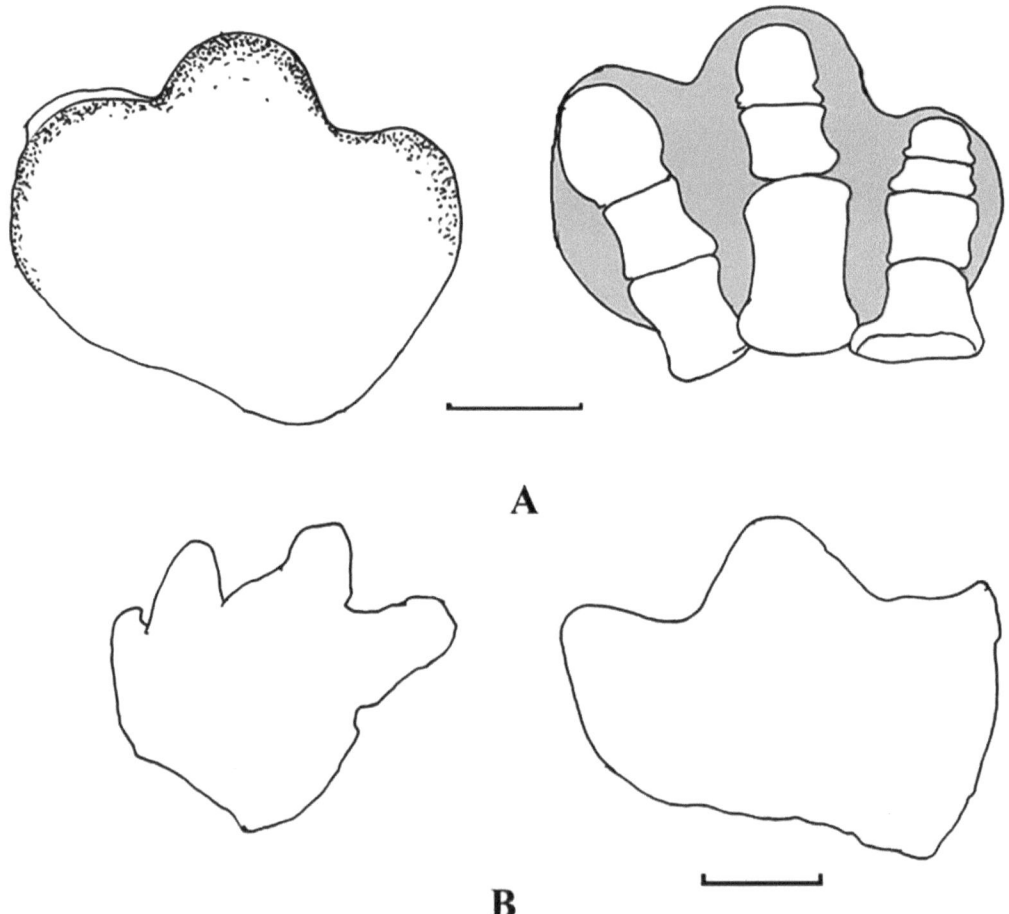

Figure 3) Tracks of *Stegopodus czerkasi*; A) Left CEUM 20571, right superimposed pes into track; B) left manus of former-holotype CU-MWC 195.1, and right, present holotype left pes (CU-MW 195.2). Scale 10 cm.

In Bakker's well-read book, Dinosaur Heresies (1986) he wrote about stegosaurs feeding in a tripodal stance. In his article that was published in the Dinosaurs Past and Present (volume 1, 1987) he elaborated more on the tripodal feeding and also pointed out that *Stegosaurus* had several bipedal characteristics (Figure 4). He followed not only Marsh but also the English biophysicist D'Arcy Thompson (1885) in which the tail vertebrae, along with the back muscles and ligaments acted like a suspension bridge. This would have helped *Stegosaurus* rear up on its hind legs and rest on its tail while feeding on higher plant material. The shape of the caudal vertebrae, and chevrons allowed for large muscles that would have easily helped in not only walking bipedal but also in a tripodal feeding stance. The dorsal vertebrae had very elongated arches that raised the rib heads to the same level as the summit of the neural spine. The dorsal vertebrae were not held rigidly with ossified tendons and the zygapophyses were large and the dorsal column permitted extensive flexion and extensions. Though I believe that the plates themselves would have stiffened the back and tail to some degree. T. R. Karbek, 2002 (an anagram of R. T. Bakker) argued for bipedality in *Stegosaurus* and *Kentrosaurus* (Figures 5, 6). The great contrast between the front and hind legs, its body proportions indicate that the center of gravity moved toward the hips, along with the position of the plates all enhanced the agility in a bipedal stance. Also, the development of neural tissue near the propulsive hind limbs reduced time lag of neural reactions. The Polish paleoartist (Krzystof Kuchnio) came to a similar conclusion and based his findings on the HMN specimens of *Kentrosaurus aethiopicus* Henning (1915) and created a computer animation showing a bipedal walking *Kentrosaurus* skeleton. The tails were held horizontally, and this would help in walking bipedally.

Gierlinsky and Sabath bring up an interesting point. One of the problems of a bipedal stegosaur is the large (not in length but girth) front legs. The humerus has a huge humeral crest, the ulna and radius are also very thick. The claws are dorsalventrally flattened and very wide. They look like a spade and I believe they could easily dig with their front feet, which might explain the thick and muscular front legs.

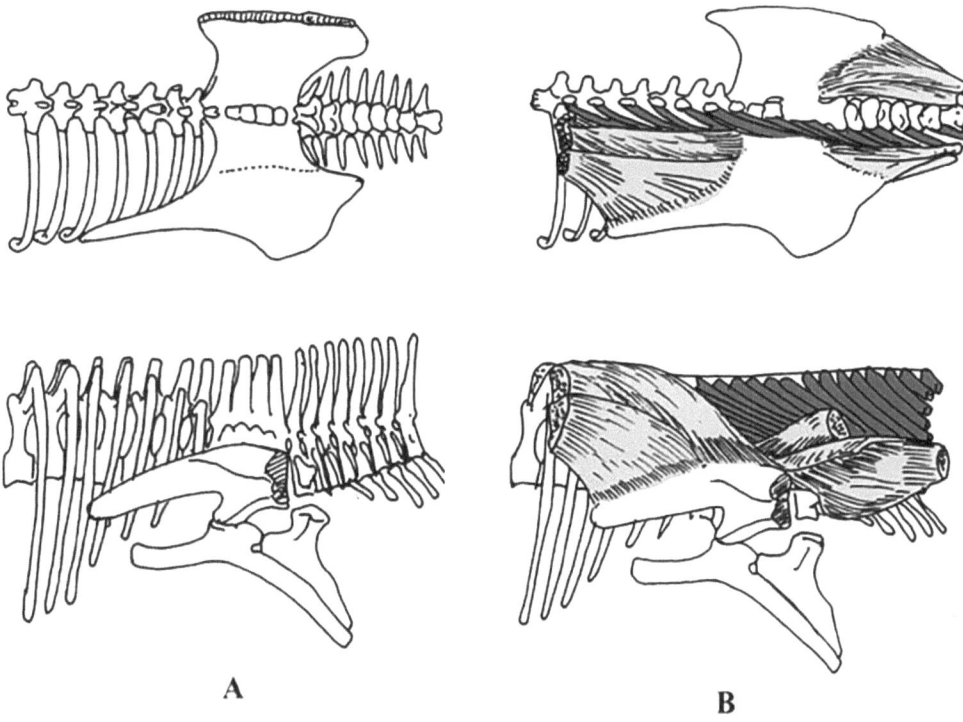

Figure 4) Trunk, sacrum and caudal vertebrae of *Stegosaurus ungulatus*; A) skeleton, B) muscles (modified from Bakker, 1987)

Did stegosaurs walk bipedally? It is possible, as the Spanish tracks show, but they also walked quadrupedally, as the North American tracks indicate. So be creative and illustrate bipedal and quadrupedal stegosaurs. But that is not to say all stegosaurs had longer hind legs than front legs. Some had more evenly matched legs (Figure 7) and didn't walk on their hind legs; i.e. *Dacentrurus* and *Huayangosaurus*

Bibliography

Bakker, Robert T. 1986. The Dinosaur Heresies, New Theories Unlocking the Mystery of the Dinosaurs and their extinction. William Morrow and Company, Inc. New York: 481pp.

Bakker, R. T., 1987, The return of the Dancing Dinosaurs: In: Dinosaurs Past and Present, v. 1, p. 38-69.

Gierlinski, G., and Sabath, K., 2002, A probable stegosaurian track from the Late Jurassic of Poland: Acta Palaeontologica Polonica, v. 47, n. 3, p. 561-564.

Gilmore, C. W., 1914, Osteology of the armored dinosauria in the U. S. National Museum with special reference to the genus *Stegosaurus*: Bulletin of the United States National Museum, v. 89, p.1-140.

Henning, E., 1915, *Kentrosaruus aethiopicus*, der Stegosauridae des Tendaguru: Sitzungsberichte Naturforschender Freunde zu Berlin, 1915, n. 6, p. 219-247,

Henning, E., 1916, Zweite Mittelung uber den Stegosauriden vom Tendaguru Sitzungsberichte Naturforschender Freunde zu Berlin, 1915, p. 175-182.

Henning, E., 1925, *Kentruruosaurus aethiopicus*, die Stegosaurier- Funde von Tendaguru, Deutsch-Ostafrika: Palaeontologrpahica Supplement v. 7, n. 1.1, p. 101-254.

Karbek, T. R., 2002, The case for *Stegosaurus* as an agle, cursorial biped: Journal of Vertebrate Paleontology, v. 22, supplement to n. 3, Abstracts of Papers, Sixty-second annual Meeting of the Society of Vertebrate Paleontology, Sam Noble Oklahoma Museum of Natural History, University of Oklahoma, Norman Oklahoma, October 9-12, p. 73a.

Lockley, M. G., and Hunt, A. P., 1998, A probable stegosaur track from the Morrison Formation of Utah: In: The Upper Jurassic Morrison Formation: An Interdisciplinary Study, Denver Museum of Natural History, Denver USA, May 26-28, 1994. Edited by Carpenter, K., Chure, D. J., and Kirkland, J. I., Modern Geology, v. 23, part 2, p. 331-342.

Maidment, S. C. R., Norman, D. B., Barrett, P. M., and Upchurch, P., 2008, Systematics and phylogeny of stegosauria (Dinosauria: Ornithschia): Journal of Systematic Palaeontology, v. 6, n. 4, p. 367-407.

Sarjeant, W. A. S., 2001, Dinosaurs in fiction: In: Mesozoic Vertebrate Life, edited by Tanke, D. H., and Carpenter, K., Indiana University Press, p. 504-529.

Thompson, D'A. W., 1885, On the systematic position of the Chameleon and its Affinities with the Dinosauria: Nature, p. 562-563.

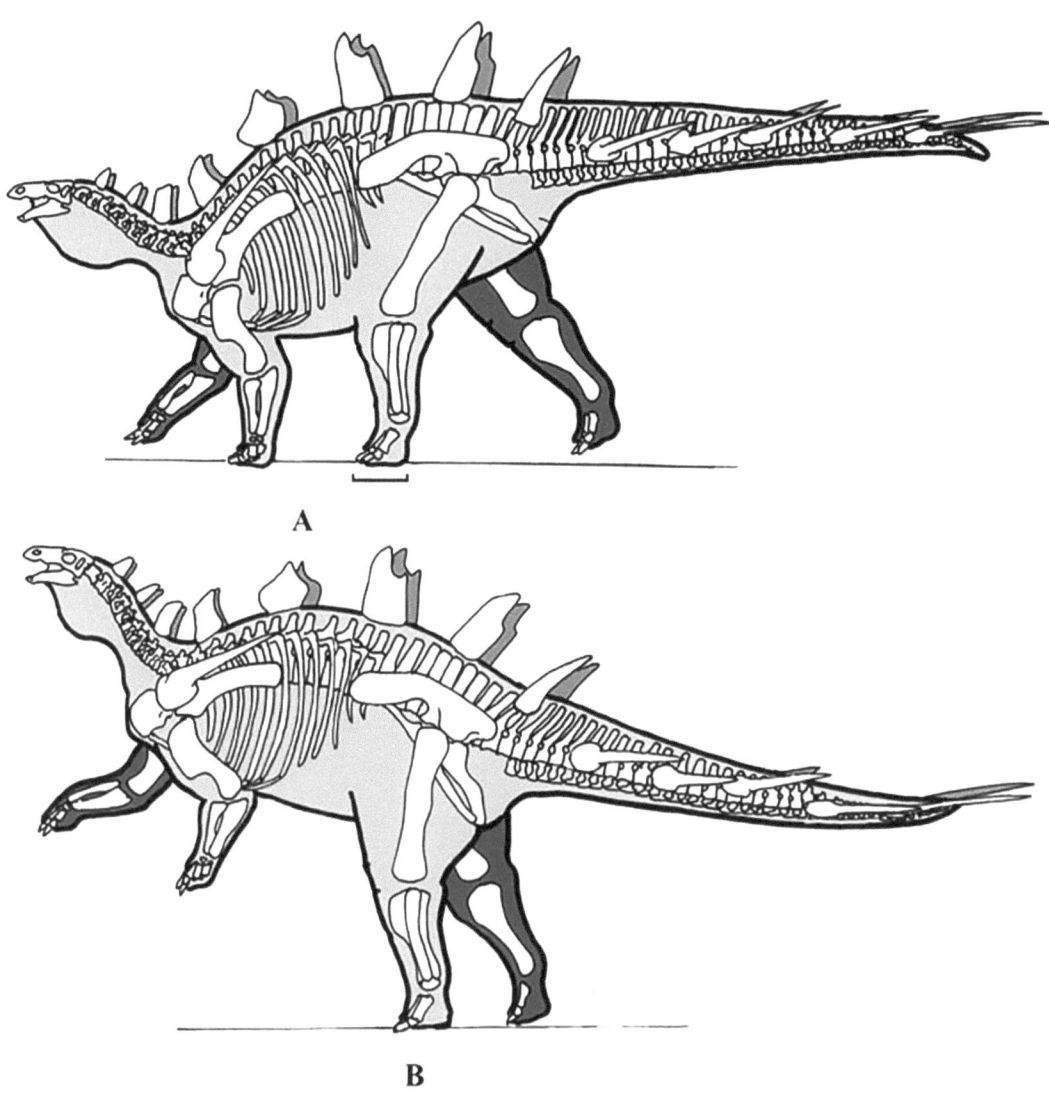

Figure 5) *Kentrosaurus* in A) quadrupedal pose; B) bipedal pose (modified from Kuchnio). Scale 20 cm.

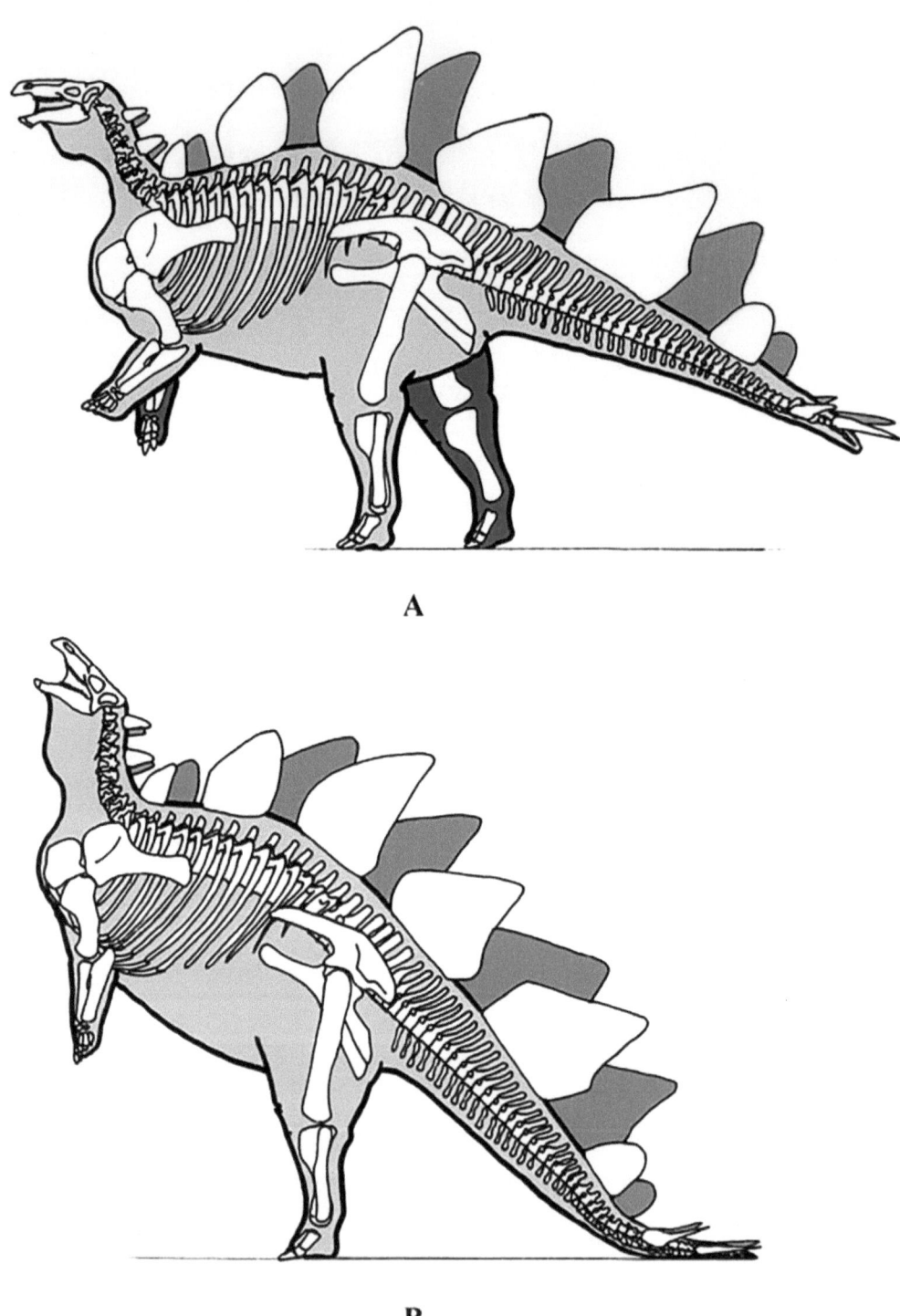

Figure 6) *Stegosaurus* in A) tripodal position (modified from Bakker); bipedal pose (Modified form Kuchnio).

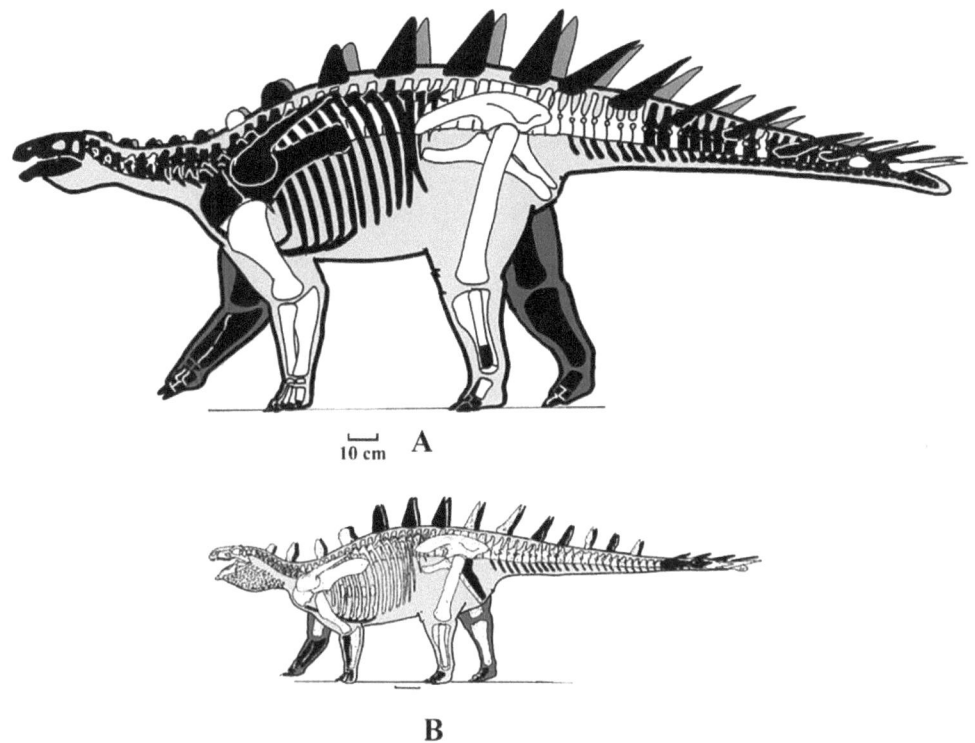

Figure 7) Skeleton of Stegosaurids that had more evenly proportioned hind legs; A) *Dacentrurus armatus*; B) *Huayangosaurus taibaii*. Scale 20 cm.

Ford, T. L., 2009, How to Draw Dinosaurs, Face Fighting Ceratopians. Prehistoric Times, n. 89, p. 18-19.

Chapter 17

Face Fighting Ceratopians

The Ceratopian Volume won't be out until the next SVP or around October 2009, so I won't be writing about the possibility of aquatic *Psittacosaurus* until the volume is published. This issue I thought I'd write about a newly published paper about fighting Ceratopians and I'll also be making a few comments about *Styracosaurus* itself.

Herbivorous dinosaurs are thought to have had a peaceful non-aggressive lifestyle and it was the predatory dinosaurs that were aggressive. For example, the fighting dinosaurs (*Velociraptor* vs *Protoceratops*) it is the *Velociraptor* that is always thought to have been the aggressor, though I believe it was the *Protoceratops*. Farke, Wolff and Tanke research shows that ceratopians weren't the peaceful animals they are generally believed to be (Evidence of Combat in *Triceratops*, PLOS One, freely available at http://www.plosone.org/article/info:doi/10.1371/journal.pone.0004252). Their study was on the pathologies on the skull of ceratopians and how they could have been made. These pathologies indicate aggressive behavior between rivals, ether from mating behavior or fighting. Farke first gave a talk about this at the 2004 SVP. The horns and frill of ceratopians are believed to either be for display, sexual dimorphism or combat.

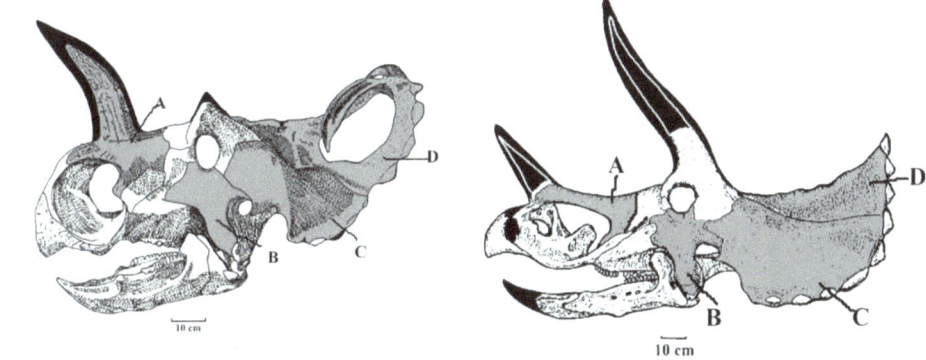

Figure 1) Diagram showing the skull bones used in the study by Farke, et al. *Centrosaurus* and *Triceratops*. A) Nasal (0/41 in *Centrosaurus* and 0/47 in *Triceratops*); B) Squamosal (5/56 in *Centrosaurus* and 7/39 in Triceratops); C) Squamosal (1/62 in *Centrosaurus* and 10/58 in *Triceratops*); D) Parietal (2/62 in *Centrosaurus* and 1/45 in *Triceratops*).

Ceratopians have been found with pathologies on skulls and skeletons and were caused by either fighting between rivals or predator. A classic dinosaur fight is between *Tyrannosaurs rex* and *Triceratops*. There are several specimens that show that this isn't just a Hollywood fantasy. Tooth marks on the orbital horns and bones of *Triceratops* from a *Tyrannosaurus rex* have been found, as well as one (or possible more) specimens that actually have an orbital horn bitten off!. In fact, one specimen of *Triceratops* specimen indicates that a *Tyrannosaurus* literally bit its carcass in two! It came in from the belly region and there are tooth marks at the sacral/dorsal vertebrae contact which indicates it cut the carcass in half! The amount of pressure needed to do that is staggering.

The authors studied adult *Triceratops* and *Centrosaurus* specimens to see if they not only showed evidence of head fighting, but also to see if there is an indication of similar fighting styles between the two. Morphologically they are very different, *Triceratops* has two large horns over its eyes and a small one on the nose, while *Centrosaurus* has a large horn on the nose and two small (or in some specimens, non-existent horns) over the eyes. Their research shows that the marks (lesions) on the skull (nasal, jugal, squamosal and parietal bones) are different which indicates they fought in different ways (Figure 1).

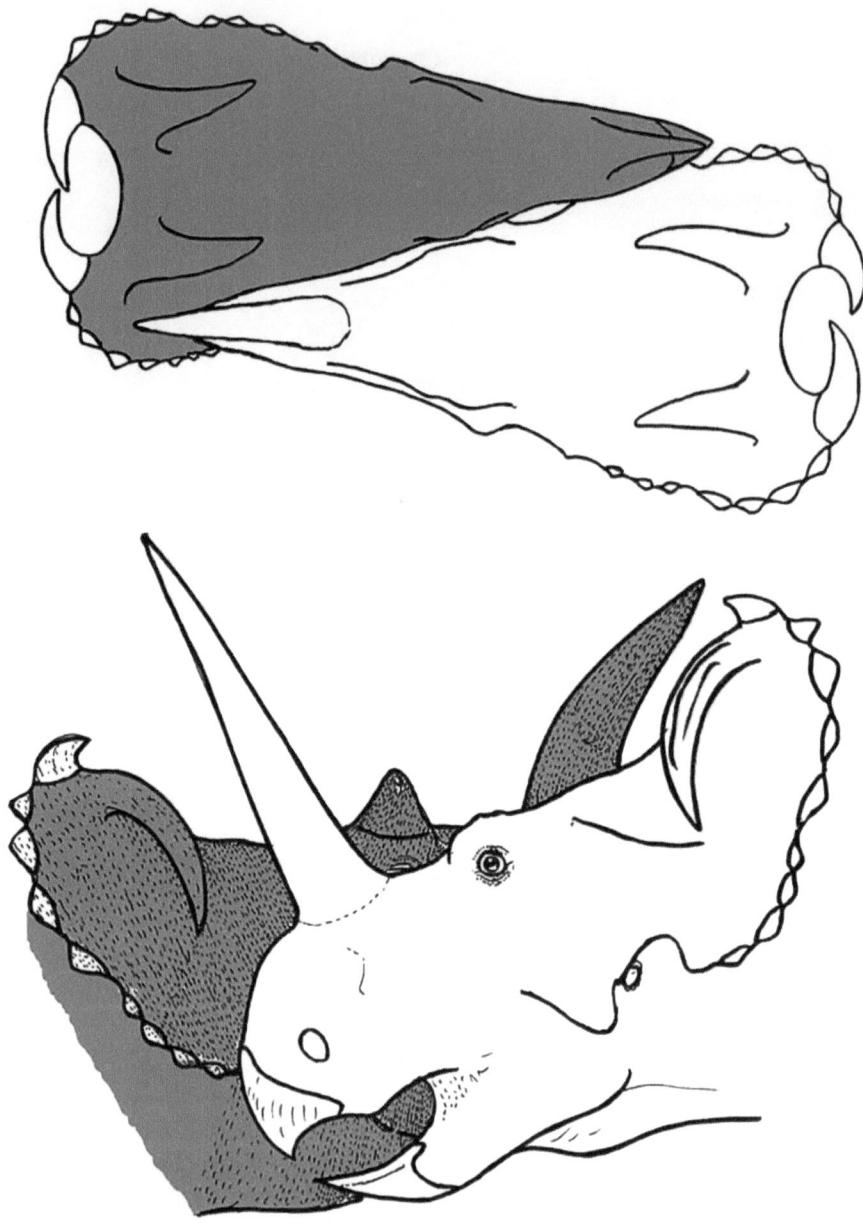

Figure 2) *Centrosaurus* fighting in dorsal and lateral view.

There are two types of pathologies that they observed, periosteal and calluses. Periosteal reactive bone is an indication of a traumatic event. It is caused by the periosteum from underlying layers of bone from subsequent inflammatory response to the bone due to the natural healing process. The bones show a remodeled ridge on the external surface which may cut across the normal pattern of neurovascular impression on the surface of the skull. They noticed this remodeling in 12 of the 26 observed lesions. Calluses are healed or healing fractures and result from several steps of bone growth. A primary callus with disorganized bone due to a secondary callus of secondary bone.

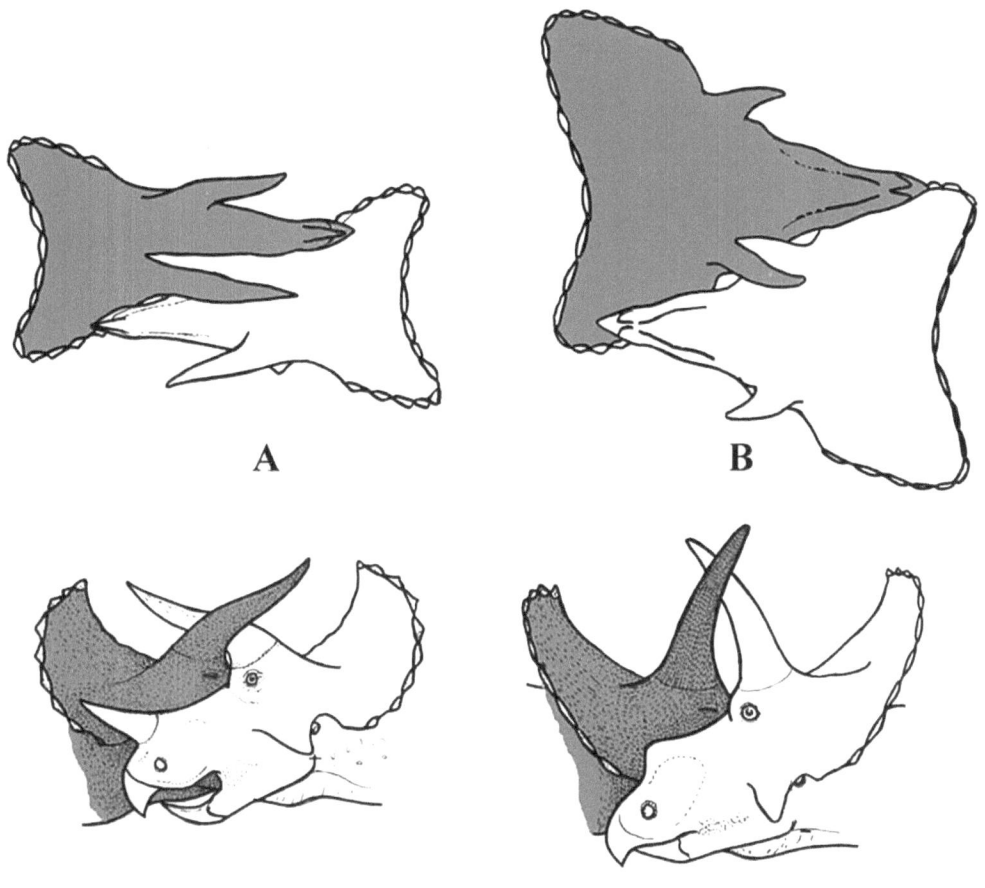

Figure 3) *Triceratops* (A) and *Nedoceratops* (B) fighting in dorsal and lateral view.

Triceratops had more lesions on the squamosal than *Centrosaurus*. Not surprising since *Triceratops* has a larger squamosal. *Centrosaurus* would have placed their heads side to side, from predentary to orbital area and probably pushed against each other using side to side action (Figure 2). The amount of pressure needed to make the lesions on the skull bones is extreme. Because of this they must have had scars, cuts and burses on their face. *Triceratops* would have done the same, but also interlocked their orbital horns which is indicated by the pathologic skull bones. *Triceratops* possibly engaged in shoving matches like that seen in turtles and unicorn beetles (Figure 3). It may have been a noisy match, snorting, grunting, and pushing against one another. I had thought the *Triceratops/Nedoceratops* specimens with vertical horns couldn't have locked horns, but after doing the illustration I could see that they could have. *Triceratops* with vertical horns are; *T. hatcheri* (*Nedoceratops hatcheri*) or *Triceratops albertensis*. *Diceratops* was shown to be preoccupied by an insect and renamed by Matues (2008) to *Diceratus*, but that turned out to be after it was renamed *Nedoceratops* a year earlier by Ukrainsky, 2007. The proper name of *Diceratops* is now *Nedoceratops hatcheri*.

Styracosaurus is a very recognizable dinosaur. The horns on the back of the frill are unique to this bizarre looking animal. Only a few specimens are known, the type *Styracosaurus albertensis*, *Styracosaurus parksi* and *Styracosaurus ovatus* as well as from a bone bed at Dinosaur Provincial Park in Alberta, Canada. *Styracosaurus ovatus* is known from the Two Medicine Formation of Montana and will be given its own genus name in the Ceratopian volume. *Styracosaurus parksi* is believed to be another specimen of *Styracosaurus albertensis*. *Styracosaurus parksi* skull is known from only a few elements, including a partial frill. A *Styracosaurus* bone bed is known from Dinosaur Provincial Park (Many years ago Darren Tanke took me to the *Styracosaurus* bone bed, and in a PaleoWorld episode both of us were filmed at that site). Interestingly the skull of *Styracosaurus albertensis* is pathologic. Holmes, Ryan and Lloyd gave a talk at the 2007 SVP, and Holmes and Ryan the same year published a redescription of the skull of *Styracosaurus*. They document the posterior edge of the frill had collapsed down, between horn 2 and horn 3, and these two horns are closer to each other than they should be (Figure 4). Holmes, Ryan and I also believe the skull is dorsally-ventrally flattened. The skull looks flatter than it should be, the frill is nearly at the same level as the skull roof; the orbit is off set. I've made a new interpretation of the skull (Figure 5). It was thought that *Styracosaurus* and *Centrosaurus* are the same genus, but *Centrosaurus* and *Styracosaurus* never lived at the same time. *Centrosaurus* is known from the Lower Dinosaur Park Formation and *Styracosaurus* replaced *Centrosaurus* in the Upper Dinosaur Park Formation.

Bibliography

Farke, A. A., Wolff, E. D. S., and Tanke, D. H., 2009, Evidence of combat in *Triceratops*: PLOS One, v. 4, Issue 1, 4pp.

Holmes, R. B., Ryan, M. J., and Lloyd, D., 2007, Restoration of the pathological parietal ornamentation of the holotype skull of *Styracosaurus albertensis* (CMN 344) with reference to new undistorted specimens: In: Ceratopsian Symposium, Short Papers, Abstracts, and Programs, complied by Braman, D. R., p. 78-79.

Holmes, R. B., Ryan, M. J., and Murray, A. M., 2006, Photographic atlas of the postcranial skeleton of the type specimen of *Styracosaurus albertensis* with additional isolated cranial elements from Alberta: Syllogeus, n. 75, 75pp.

Mateus, O, 2008, Two ornithischian dinosaurs renamed: *Microceratops* Bohlin 1953 and *Diceratops* Lull 1905: Journal of Paleontology, v. 82, n. 2, p. 423.

Ukrainsky, A. S., 2007, A new replacement name for *Diceratops* Lull, 1905 (Reptilia: Ornithischia: Ceratopsidae): Zoosystematica Rossica, v. 16, n. 2, p. 292.

Ukrainsky, A. S., 2009, Synonymy of the Genera *Nedoceratops* Ukrainsky, 2007 and *Diceratus* Mateus, 2008 (Reptilia: Ornithischia: Ceratopidae): Palaeontological Journal v. 43, n. 1, p. 116.

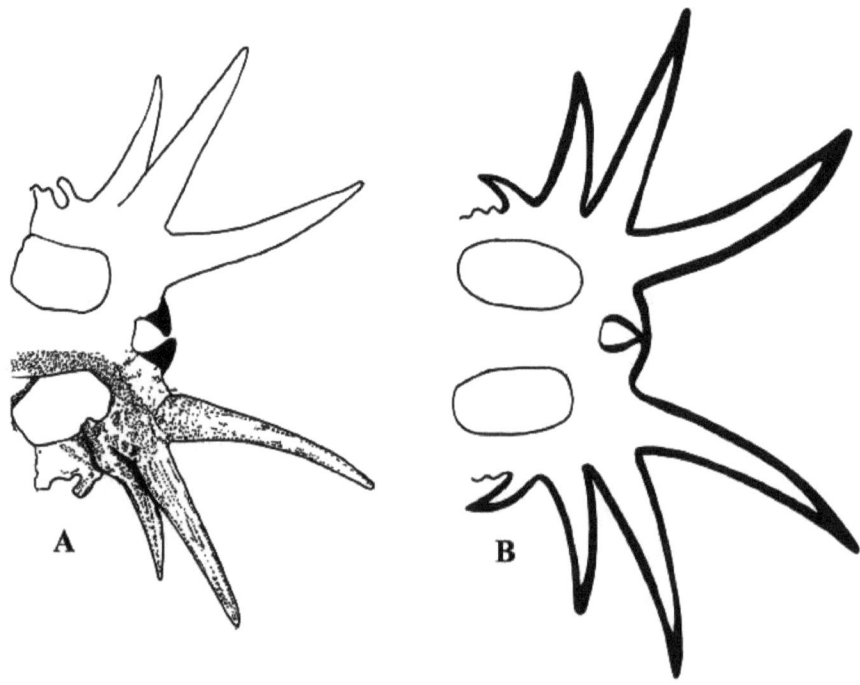

Figure 4) *Styracosaurus* frill showing pathology (A) and new interpretation (B).

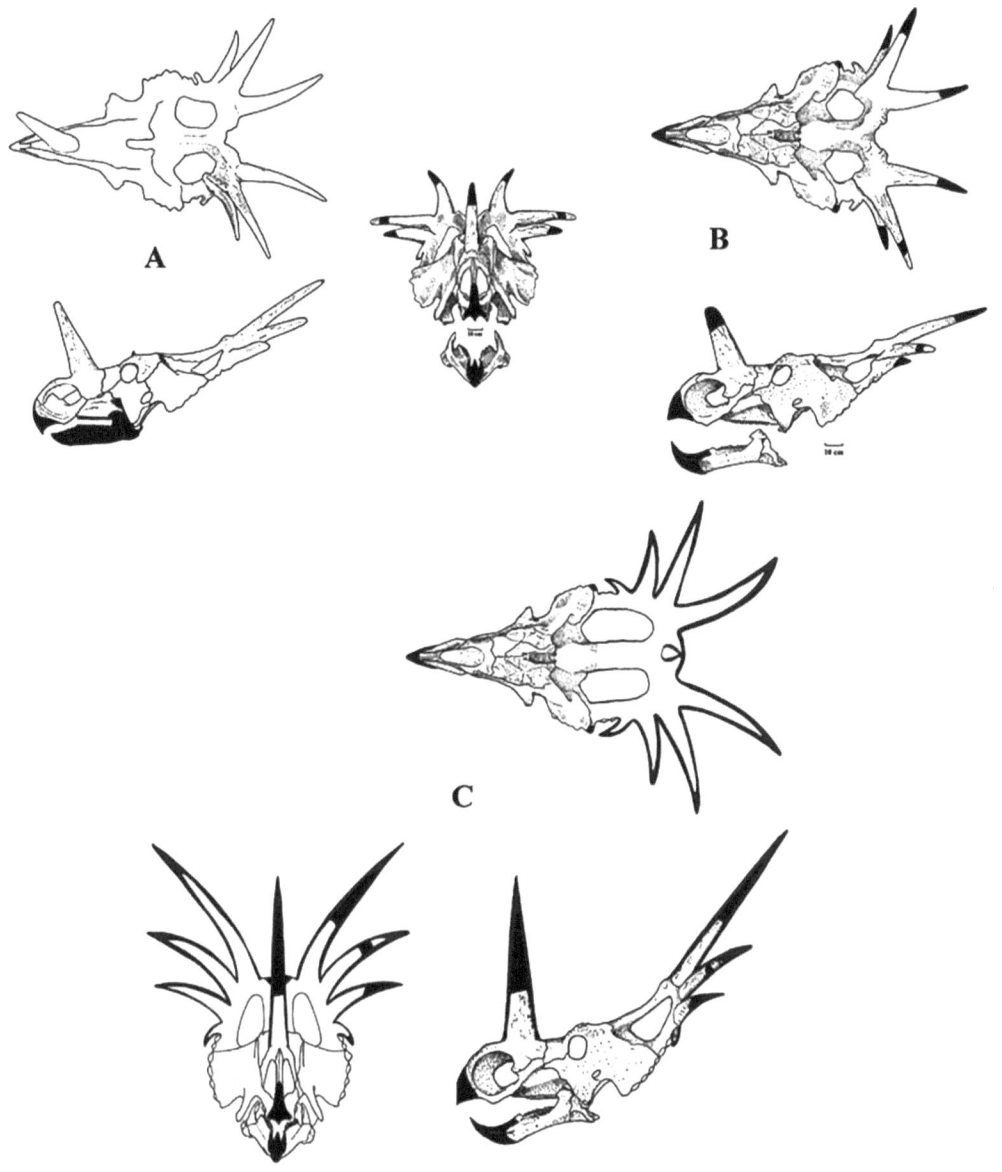

Figure 5) *Styracosaurus albertensis* skull, A) after Lambe, B) after Holmes and Ryan, C) new interpretation of the skull.

Ford, T. L., 2009, How to Draw Dinosaurs, The dinosaurs with a bad name...Oviraptorids, Part 1. Prehistoric Times, n. 90, p. 18-19.

Chapter 18

The dinosaurs with a bad name...Oviraptorids, Part 1

This article stems from a query that Greg Paul gave me months ago. He asked if I had written any PT articles about Oviraptorids, of which I told him I had not. So, I thought the next article/s would be about Oviraptorids or more correctly Oviraptorosauria. There's a lot to talk about so I'll split this into two articles. The first will be on the skull, the second on the skeleton. Oviraptorosauria range in size from small animals (*Caudipteryx, Protarchaeopteryx*, which are about a meter long) to very large (*Gigantoraptor*, 8 meters long and 3 meters tall). Some have been found with feathers, and I'll be addressing that in part 2. Their systematic position in theropoda is controversial. Some believe they are true advanced theropods, others believe they are actually birds.

Oviraptorosauria are edentulous theropods with strange looking skulls and skeletons. Some have long skulls, others short, some have crests others do not. Even the morphology of the hands differs with some with long fingers and some with short. *Caudipteryx* and *Protarchaeopteryx* [= *Incisivosaurus*], being the exception with just premaxillary teeth in *Caudipteryx* and large premaxillary and smaller maxilla and dentary teeth in *Protarchaeopteryx*. There is some confusion on which skull goes with which name. Hopefully I'll fix that with this article.

The first officially named Oviraptorosauria is obviously *Oviraptor philoceratops*. It was found in the Red Beds from the famous Flaming Cliffs of Mongolia. The American Museum of Natural History's expeditions discovered the fragmentary specimen. It was found 'sitting' over a nest of eggs. It was thought to have been 'stealing' the eggs from *Protoceratops*, which is a more abundant dinosaur in the area, hence the name 'Egg Thief'. It wasn't until the early 2000's when it was found out that it wasn't stealing the eggs of *Protoceratops* but was protecting its own eggs.

The skull of Oviraptorosauria are unique among the theropoda. They have a solid premaxilla, and nearly non existent strap-like maxilla. The skull itself is lightly built. From the orbit back, the skull dip downward and dose the opposite from mid section of the skull to the beak. The premaxilla has ridges and bumps on the lower edge and more than likely had supported a keratinous covering. All Oviraptorosauria have large orbits with the lower jaw being the widest just in front of the orbit. Research has shown that the inner ear in Oviraptorosauria was similar to birds and advanced theropods including, *Archaeopteryx*, dromaeosaurids and troodontids. Their large eyes may indicate a nocturnal lifestyle and the enhanced hearing may have helped them at night.

The lower jaw in many genera has a 'bump' at its highest point. This bump on the upper lower jaw extends into the orbit when the jaw was closed. I wonder if it affected the eye itself. Its nares were small and sat high on the anterior end of the skull. The skull in many ways looks like a parrot, but unlike a parrot the 'beak' doesn't move and is solid. And like a parrot it may have had a fleshy cheek limiting the size of the exposed mouth (I had a conversation with paleoartist Berislav Krzic about this and this is what he believes) (Figure 4). Unlike the premaxilla the predentary lacks the bumps and ridges and is more smooth. The jaw muscles in Oviraptorosauria was different than other theropoda. They have a unique jaw morphology in that the lower jaw has an angle (from the middle dorsal edge to posterior end) and the jaw muscles wouldn't have had the normal vertical structure from the dentary to the supratemporal fenestra but would have angled from the dentary to the supratemporal fenestra (Figure 4). This would affect the jaw movement and there have been several studies which shows the lower jaw move both up and down and forward and back. Unlike other theropods in which the palate in concave the palate of Oviraptorosauria is convex and extends lower than the exterior margin of the upper jaw. The front of the palate (in some Oviraptorosauria) has 2 large 'teeth' and I believe that they were able to pick up eggs in its mouth and puncture the shell with these specialized teeth. I'm not saying they exclusively ate eggs, but eggs may have been part of their diet.

The crested Oviraptorosauria genera are; Genus nova (GI SPS No. 100/42), *Citipati, Nemegtomaia* (*Nemegtia*), *Rinchenia, Oviraptor* and *Chirostenotes* (Figure 1). There are two undescribed specimens that

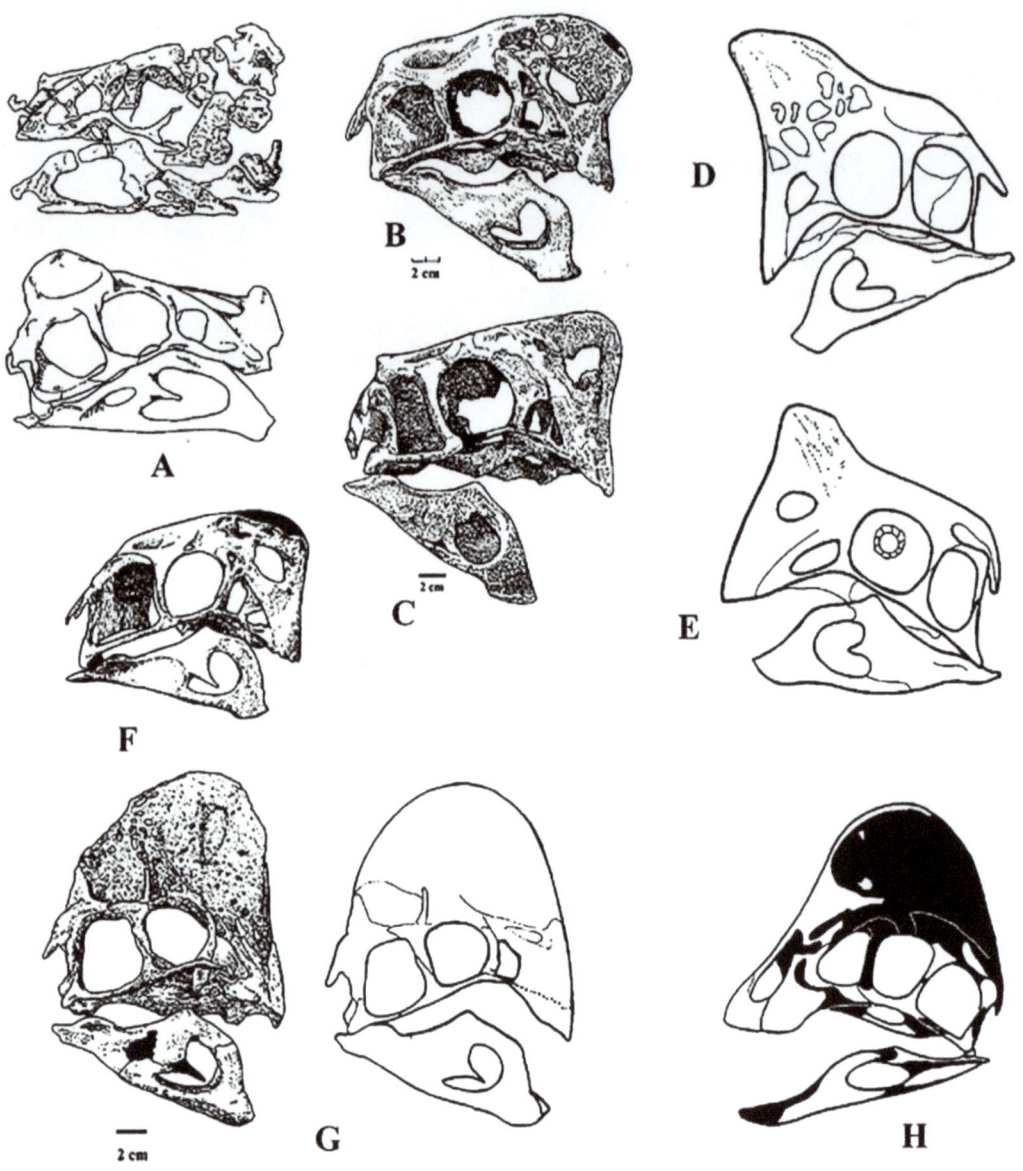

Figure 1) Crested Oviraptorosauria; A) *Oviraptor philoceratops* (AMNH 6517); B) Oviraptor nova GI SPS No. 100/42; C) *Nemegtomaia barsboldi* GIN 100121 12; D) Oviraptor nova (Gaston); E) Oviraptor nova (Gaston); F) *Citipati osmolskae* IGM 100/978; G) *Rinchenia mongoliensis* GI SPS No. 100/32 (after Barsbold, and reconstructed skull); H) *Chirostenotes sp* (Triebold specimen, after Scott Hartman)

casts are being sold of that has a high short 'pointed' crest. Rob Gaston is selling them and I have permission to illustrate the specimens from Rob himself (http://www.gastondesign.com/aboutus.htm). The crests are thin and highly pneumatic and may have supported a keratinous covering. The shape of the crest differs from genera to genera. Some have the highest point near the front of the beak, others have the highest point above the orbit similar to that of a Cassowary. Unfortunately, *Oviraptor's* skull isn't well preserved and its shape is not certain. It had been theorized after it was described, that it had a small horn, which is probably just a fragment of the crest. *Oviraptor* and *Chirostenotes* have the longest skulls of Oviraptorosauria.

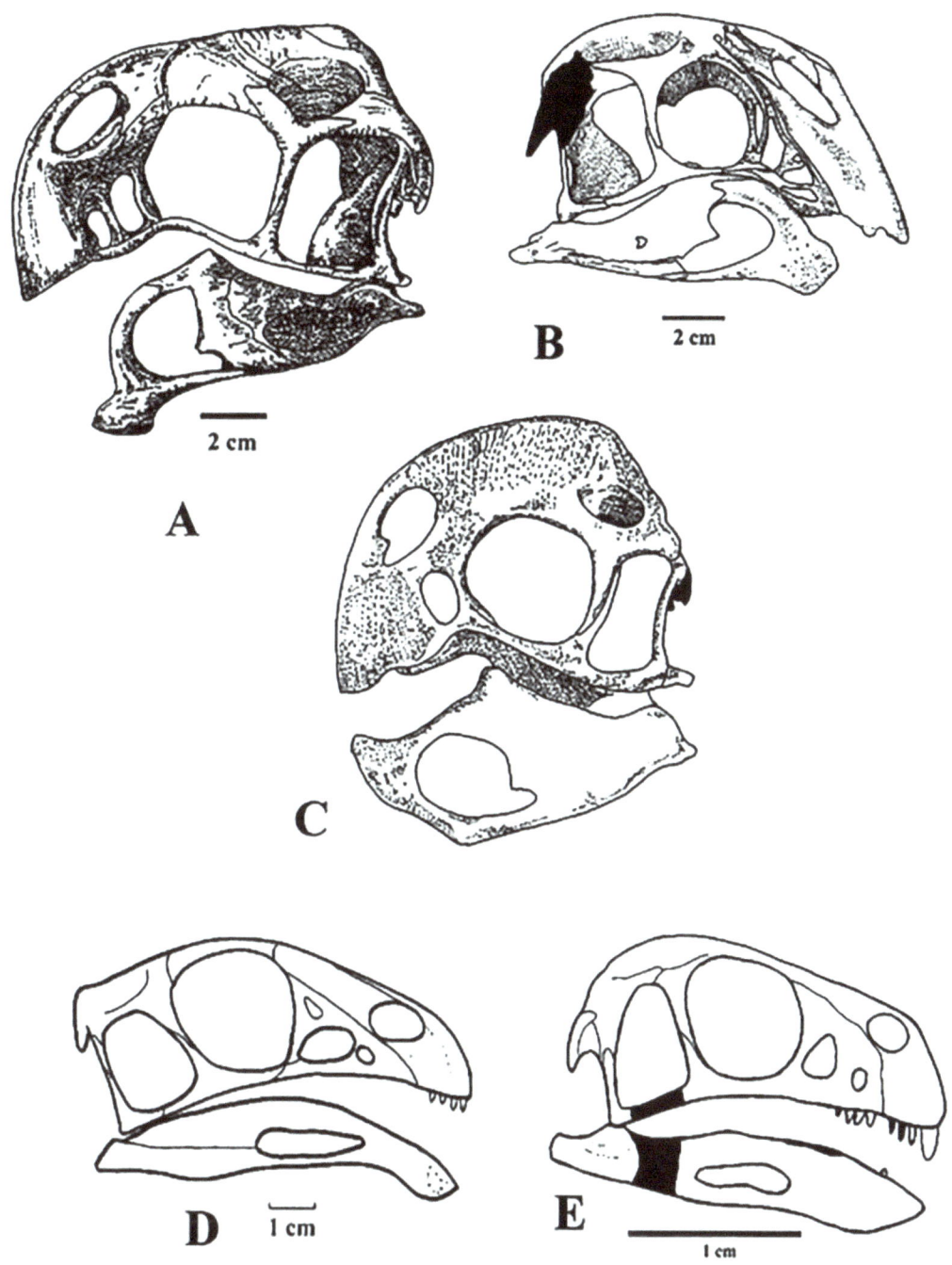

Figure 2) Crestless Oviraptorosauria; A) *Conchoraptor gracilis* GKH PST No. 100/20; B) *Khaan mckennae* IGM 100/1127; C) *Ingenia yanshini* GI SPS No. 100/30; D) *Caudipteryx zoui* NGMC 97-4-A & NGMC 97-9-A; E) *Protarchaeopteryx robusta* = *Incisivosaurus gauthieri* IVPP V13326.

I remember being at the Tucson Rock Show about 10 years ago and Mike Triebold showed me some casts of the skull elements of *Chirostenotes* and asked if I knew how they went together (you can visit Mike Triebold's site at http://www.trieboldpaleontology.com or at http://www.rmdrc.com/index.htm). It was a daunting task which I wasn't able to perform, but after research and diligence Mike came up with what was probably what it really looked like. It had a very large, oval crest. *Citipatia*, *Nemegtomia* and

100/42 have the shortest crests. *Rinchenia* and *Chirostenotes* have the largest (tallest) crest. The skull of *Oviraptor philoceratops* is longer than the other known Asian Oviraptorosauria, and after looking at the reconstruction of *Chirostenotes* skull I believe that the skull of *Oviraptor philoceratops* may have looked more like *Chirostenotes* than the other 'shorter' skulled Oviraptorosauria. Though the skulls do look similar their skeletons are very distinctive and only look superficially alike (Phil Currie pers comm.).

The 'crest-less' Oviraptorosauria genera are *Caudipteryx*, *Protarchaeopteryx*, *Conchoraptor*, *Ingenia*, and *Khaan* (Figure 2). It was actually *Ingenia* that Greg Paul was asking me about. He wanted to know what the skull looked like. Its skull hasn't really been described or figured. It has a short rounded skull. All the crestless forms have short skulls.

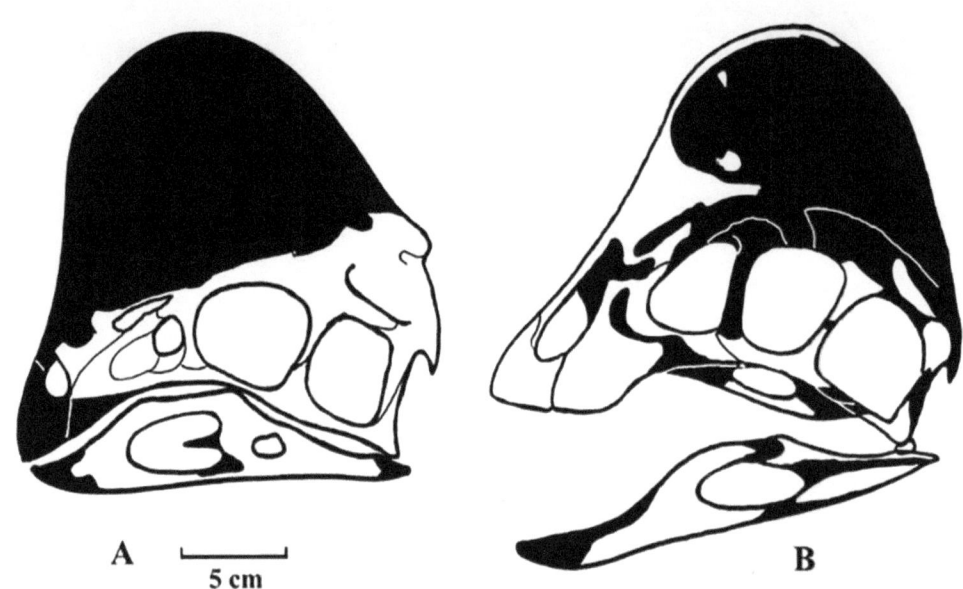

Figure 3) New interpretation of Oviraptor using *Chirostenotes* as a template; A) *Oviraptor*; B) *Chirostenotes*.

Bibliography

Barsbold, R., 1986, [Raubdinosaurier Oviraptoren]: In: Vorobyeva, E. I. (ed.). Herpetologische Untersuchungen in der Mongolischen Volksrepublik. Akad. Nauk S.S.S.R. Inst. Evolyucionnoy Morfologii i Ekologii Zhivotnykh im. A. M. Severtsova, Moskva, p. 210-223.

Barsbold, R., 1997, Oviraptorosauria: In: Encyclopedia of Dinosaurs, edited by Currie P. J., and Padian K., Academic Press, p. 505-509.

Clark, J. M., Norell, M. A., and Barsbold, R., 2001, Two new oviraptorids (Theropoda: Oviraptorosauria), Upper Cretaceous Djadokhta Formation, Ukhaa Tolgod, Mongolia: Journal of Vertebrate Paleontology, v. 21, n. 2, p. 209-213.

Ji, Q., Currie P. J., Norell M. A., and Ji S.-A., 1998, Two feathered dinosaurs from northeastern China: Nature, v. 393, p. 753-761.

Ji, Q., and Ji S.-A., 1997, *Protarchaeopteryx*, a new genus of Archaeopterygidae in China: Chinese Geology, 1997, v. 3, n. 238, p.38-41.

Lu, J.-C., Tomida Y., Azuma Y., Dong, Z.-M., and Lee, Y.-N. 2005, *Nemegtomaia* gen. nov., a replacement name for the oviraptorosaurian dinosaur *Nemegtia* Lü et al., 2004, a preoccupied name: Bulletin of the National Science Museum, Tokyo, Series C, v. 31, p. 51.

Osborn, H. F., 1924, Three new Theropoda, *Protoceratops* Zone, Central Mongolia: American Museum Novitiates, n. 144, p. 1-12.

Xu, X., Cheng, Y.-N., Wang, X.-L., and Chang, C.-N., 2002, An unusual oviraptorosaurian dinosaur from China: Nature, v. 419, p. 291-293.

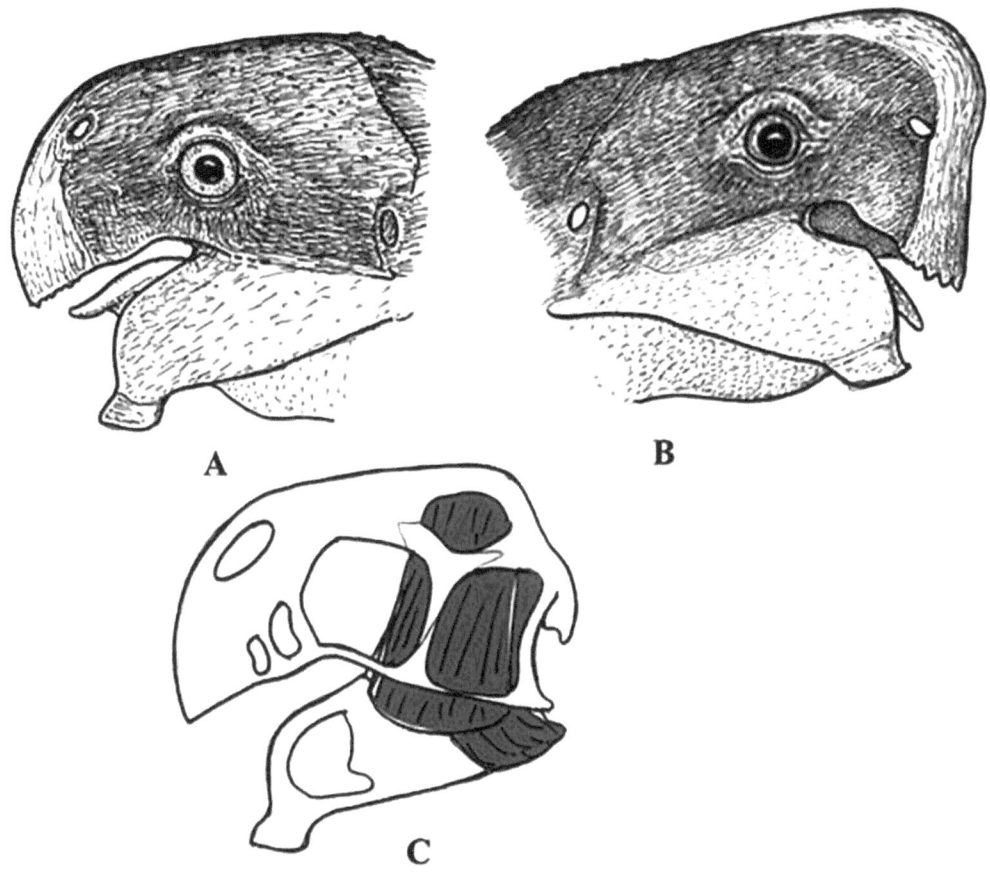

Figure 4) Head reconstructions of Oviraptorosauria; A) *Conchoraptor gracilis*; B) *Citipati osmolskae*; C) Jaw muscles of *Conchoraptor gracilis*.

Ford, T. L., 2009, How to Draw Dinosaurs, The dinosaurs with a bad name...Oviraptorids, Part 2. Prehistoric Times, n. 91, p. 18-19.

Chapter 19

The dinosaurs with a bad name...Oviraptorids, part 2

As I stated last issue in the first part of this article, the original *Oviraptor* specimen was found near a nest of eggs. The most abundant dinosaur from the Red Cliffs of Mongolia is *Protoceratops* and *Oviraptor* was believed to be 'stealing' the eggs of a *Protoceratops*. It wasn't until the late 1990's early 2000's when several brooding Oviraptorid specimens were found sitting over a nest of eggs. And even before that, the eggs from Mongolia started to get more and more scrutiny and the identity of the egg layer was challenged. It wasn't until nests and eggs were found in Montana showed that the long narrow eggs actually belonged to theropods and not ornithopods. Paleontologist compared thin sections of the dinosaur eggshells to those of birds and found that the theropod eggshells were more bird-like. Currie and Dong (1996) were the first to describe a brooding oviraptorid. Unfortunately, it was a partial specimen (oviraptorid gen sp indet). Expeditions by the American Museum of Natural History in Mongolia found the more intact brooding specimens (*Citipati*). The Oviraptorid specimens showed that theropods (at least some) did sit over nests and did protect its eggs.

Whether or not they had feathers is another question. Even though their arms were stretch over the nest doesn't necessarily mean their arms were feathered. But if not, why would they need to cover the nest, as they seem to be doing. Where they protecting the nest from the weather (i.e. sand storms, thunder storms) or from predation? By using inferences, it does seem more likely than not they were. *Caudipteryx* did have feathers on the arms, end of tail and over the body. So, would it be out of the question for artists to put feathers on Oviraptorids? No, to be blunt about, if you want to put feathers on them, then by all means go for it.

Without question, Oviraptorosaurs were strange looking theropods. The type specimen (*Oviraptor*) lacked the hind legs and tail. *Oviraptor* was depicted looking like an Ornithomimid; long neck, long legs and a long tail. I was the first to illustrate an Oviraptorosaur with a short tail (1988) (Figure 1). I was illustrating a more complete *Chirostenotes* specimen for George Olshevsky's Archosaur articulations publications. Gilles Danis discovered the specimen in the beginning of the 1979 Field Season of Dinosaur Provincial Park. Only the sacrum was uncovered and they waited for later in the field season to dig it up. George Olshevsky was part of that project and he helped lug the Jackhammer up the hill and shoved overburden away while Gilles drilled down to the bedding plane. Gilles chiseled away the matrix, gradually exposing the sacrum, an ilium, femur, tibia and some manual and pedal phalanges. George tapped away at the specimen a bit and discovered an ischium, but as soon as he told Gilles he had found some bone, he took over lest he wreck something valuable (George pers comm...). Currie and Russell published on the specimen for the first time in 1987. Their research showed there were two morphs of *Chirostenotes*, a gracile and robust morph. This is similar to the gracile and robust *Tyrannosaurus*, *Allosauurs* and *Coelophysis* specimens. The robust forms are *Chirostenotes pergracilis*, *Macrophalangia canadensis* and *Caenagnathus collinsi* (the later two have been sunk into *Chirostenotes*) and the gracile morphs are *Caenagnathus sternbergi* and *Ornithomimus elegans* (also sunk into *Chirostenotes*) and the new specimen described by Currie and Russell. I don't know where the new Triebold huge *Chirostenotes* fits in. Even though no tail had been found I drew it with a short tail because the front of the sacral vertebrae was larger/taller than the back of the sacral vertebrae. I drew a line from the top and bottom of the sacrum and drew the tail at about where the two lines converged (actually just a little longer than that because it would have been a really short tail). More complete oviraptorosaurs specimens have been found and it turns out I was right, oviraptorosaurs had short tails.

The arms and hands vary between the different genera of Oviraptorosaurs. The brooding AMNH specimen is a *Citipati* specimen, which has long arms, as does the type *Oviraptor*, and those with shorter arms are *Khaan, Ingenia, Caudipteryx*. *Citipati, Conchoraptor, Oviraptor, Chirostenotes* all have long fingers, *Citipati* and *Ingenia* have shorter fingers with *Ingenia* having the shortest fingers and largest claw on digit 1 of the hand. Like all other theropods (as it is now appearing) the palms faced each other when relaxed (Figure 2).

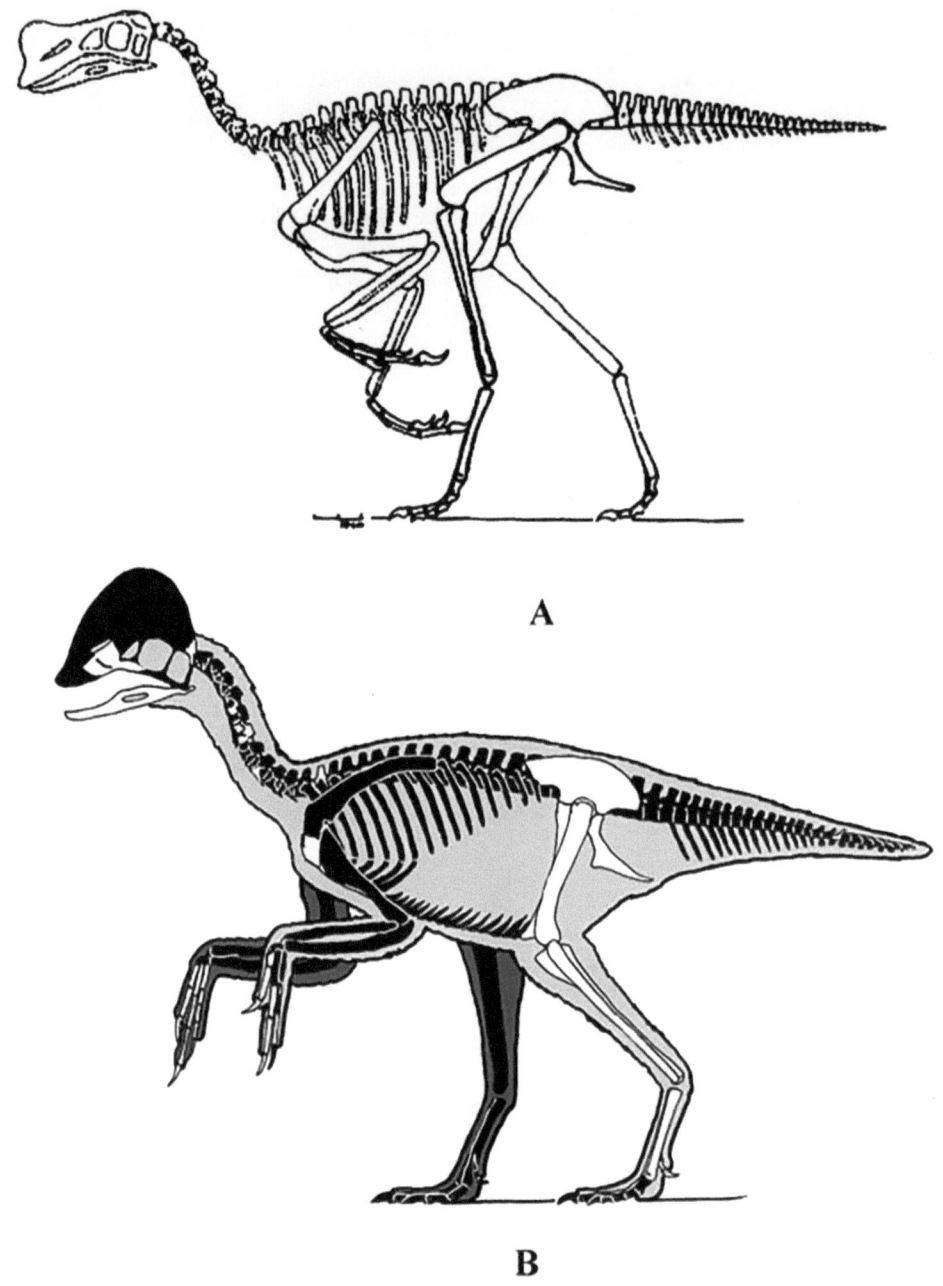

Figure 1). A) My first reconstruction of *Chirostenotes*; B) My new interpretation.

The metatarsals of oviraptorids are unfused and not like those of Ornithimimosaurids, or Tyrannosaurids. They are shorter than Ornithimimosaurs have 4 toes with 1 being the shortest and sat on the side of metatarsal 2. The skull, shoulder-girdle, forelimbs are more bird like than many other theropods and some believe to be even closer to birds than Dromaeosaurids (Figure 3).

The pelvis of oviraptorids is more similar to ornithimomimids and tyrannosaurids than dromaeosaurids in that the pubis faces forward and not parallel to the ischium (Figure 4). They have 10 cervicals, 13 dorsals 6 or 7 sacrals and about 30 or so short caudals.

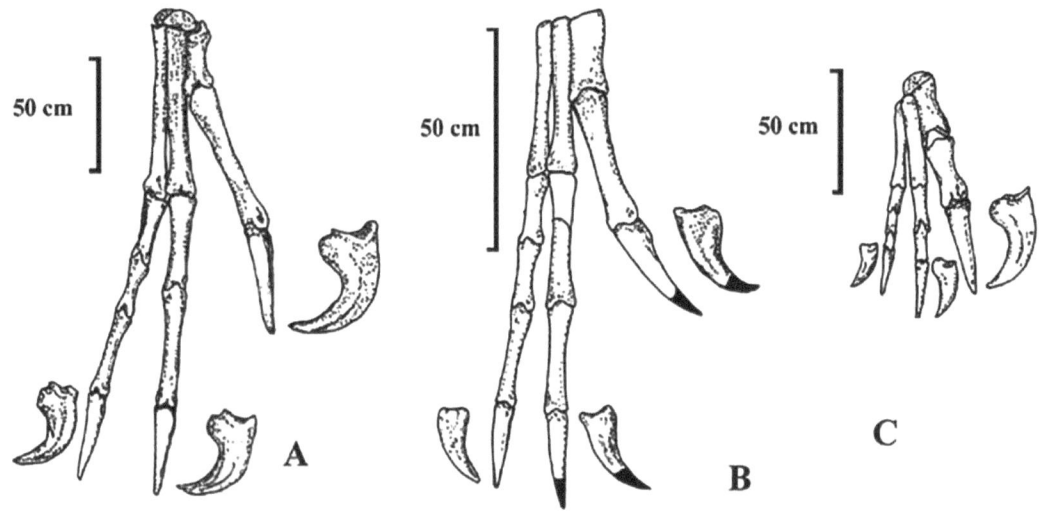

Figure 2). Hands of Oviraptorosauromorphs; A) *Oviraptor philoceratops*; B) *Conchoraptor gracilis*; C) *Ingenia yanshini*.

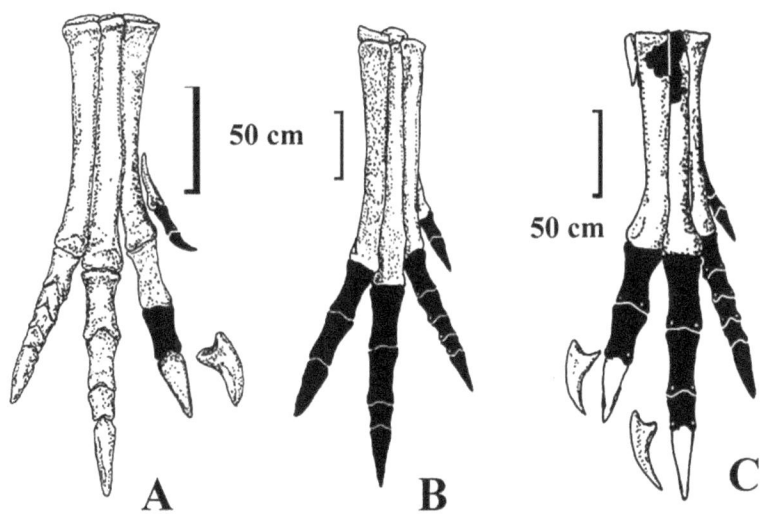

Figure 3) Feet of Oviraptorosauromorphs; A) *Ingenia yanshini*; B) *Oviraptor philoceratops*; C) *Conchoraptor gracilis*.

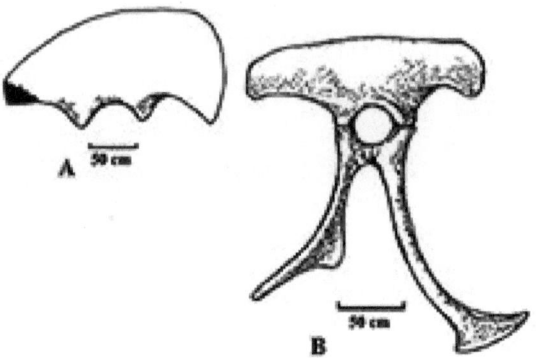

Figure 4). Pelvis and ilia of oviraptorosaurs. A) *Rinchenia* mongolensis; B) *Ingenia yanshini.*

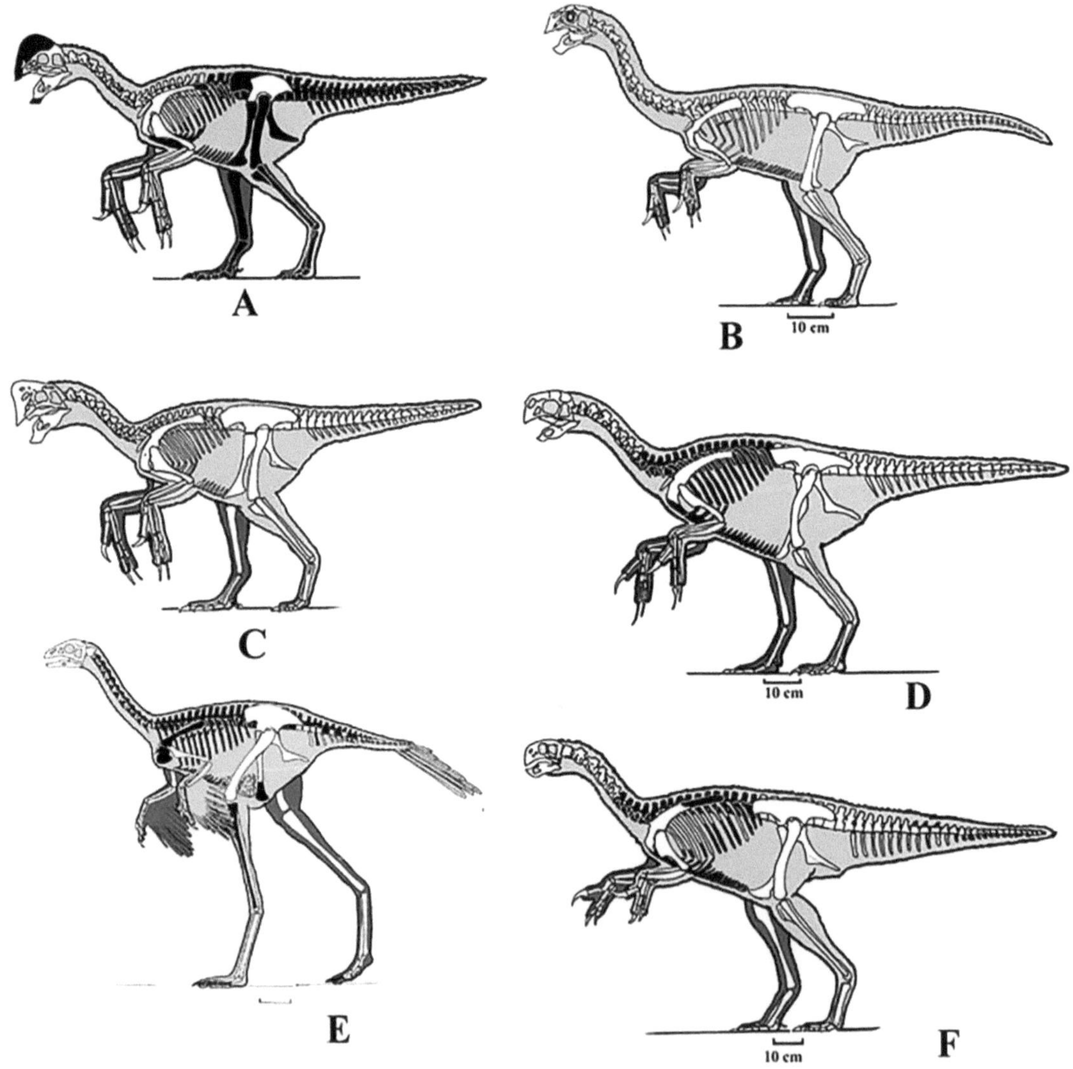

Figure 5) Oviraptorosauromorph skeletons: A) *Oviraptor philoceratops*; B) *Khaan mckennae*; C) *Citipati osmolskae*; D) *Conchoraptor gracilis*; E) *Caudipteryx zoui*; F) *Ingenia yanshini*.

Ford, T. L., 2010, How to Draw Dinosaurs, Bulking up Sauropods (and dinosaurs in general). Prehistoric Times, n. 92, p. 18-19.

Chapter 20

Bulking up Sauropods (and dinosaurs in general)

I didn't want to split up the two Ovirpatorid articles, but I believe what I'm going to be writing about here is extremely important and might very well change the look of not only sauropods but dinosaurs in general.

When I perused the Journal of Vertebrate Paleontology (issue 29, no. 2, 2009) I came across an article about the epaxial muscles of crocodiles. I immediately thought that someone should write this up for dinosaurs also, and low and behold a few articles later someone did; Daniela Schwarz-Wings (Reconstruction of the thoracic epaxial musculature of diplodocid and dicraeosaurid sauropods and was a co-author on the crocodile article). I've never been big on trying to understand what the muscle structure looked like because in dinosaurs it's somewhat subjective but am happy when someone does write about the muscles structure.

What are epaxial muscles? They are the muscles that form the bracing mechanism that accommodates the mechanical loads that act on the vertebral column during locomotion, i.e. along the neural spines and ribs. Schwarz-Wings (for the rest of the article I'll just be using Wings) used living animals as her analog; crocodiles and large birds. I'll be using the same terminology that she used. She also understands that muscle structure of sauropods my not have looked like either the birds or crocodiles she studied, but through inferences she has come up with what the muscle structure may have been like.

Where the muscles and tendons begin is called the origin and where they end is called the insertion. It is important to not only understand the muscles but also the morphology of the vertebrae.

Brake down of a vertebra; Centrum (the large spindle like structure), neural arch (just above the centrum and below the neural spine. The spinal cord runs down the middle of the neural arch and above the centrum), neural spine (above the neural spine), rugosity on the top of the neural spine (large robust top of the neural spine), spdl (spinodiapophyseal lamina) (large ridge running down the side of the neural spine), diaphyoses (two large projections on the top of the neural arch where the ribs attach), postzygophysis (where the anterior neural arch attaches to), prezygophysis (where the neural arch attaches to the posterior of the centrum). (Figure 1)

Muscle names; m ilcost (*m. ilocostalis*), m long d (*m. longissimus dorsi*), m transsp lat. (lateral part of m. *transversopinalis*), *m. transsp med* (medial part of *m. transversopinalis*).

The *ms transsp med* origin is at the top of the dorsal neural arch and connects to the opposite neural arch.

The *ms. Transp lat* origin is at the lower edge of the dorsal neural arch and insertion is to the opposite neural arch.

The *ms long d* connects the centra to each other.
As per Schwarz...

Deep layer of *m. transversospinalis* group: Origin: Base of the neural spine, cranial (anterior) face of the postzygapophyses and cranial (anterior) edges of the neural spine. Insertion: Caudal face of postygapophyses, caudal edges of neural spines.

m. transversospinalis, medial part: Origin: Tendinous or craniodorsal corner of neural spine, tendinous on caudodorsal corner of neural spine. Insertion: Fleshy on lateral faces of neural spines and adjacent tendons, roof of neural arches between zygapophyses and neural spines, zygapophyseal capsule, rugose postzygapophyseal part of spinopostzygapophyseal lamina.

m. transversopinalis, lateral part: Origin: From cranial (anterior) margins and dorsal surface of transverse processes, possibly by fascicles. Insertion: Caudal (posterior) margins of transverse processes and PODL, possibly by muscle fibers.

m. longissimus dorsi: Origin from lateralmost part transverse process, possibly by tendons: Insertion: On bounding septum to *m. transversospinals,* on lateralmost rough part of transverse processes, possibly tendinous and fleshy.

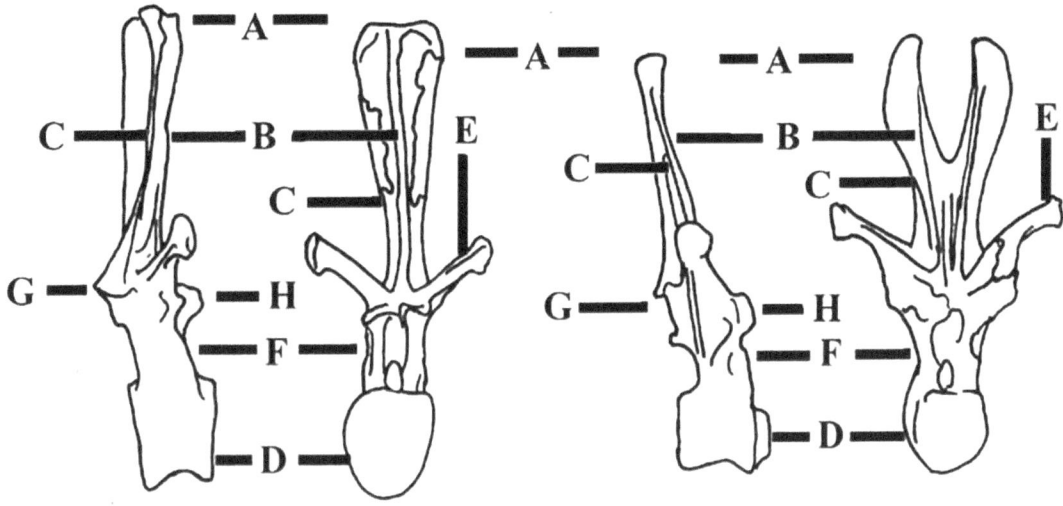

Figure 1) Diagram of vertebra of *Dicraeosaurus*; A) rugosity on the top of the neural spine; B) neural spine; C) spdl (spinodiapophyseal lamina); D) centra; E) diaphyoses; F) neural arch; G) postzygophyseis; H) prezygophyseis.

m. ilocostalis: Aponeurotically origin from preacetabular process of ilium, fleshy from medial surface of preacetabular process: Insertion: At lateral surface of dorsal ribs, possibly tendinous on caudolateral margin of dorsal ribs.

Starting at the top of the vertebral column and working down along the ribs in cross section the first muscle is the *m transsp med*, then the *m transsp lat* (*m asc th*), then the *m long d* (*m ilcost cd*) and finally the *m ilcost* (Figure 2, 3).

The deep layer of *m. transversspinals* group extends from the last cervical and continues to the caudal vertebrae. Inside this group lies tendons (as mentioned before). The *m. longissimus dorsi* extends on the lateral side of the last cervical to the caudal vertebrae and extends down to the lower middle area of the ribs. These muscles were extensive and thick and along with the skin would have had given sauropods a thick hide. The *ap ilost* origin is on the rugosity cranial (anterior) edge of the ilium and inserts on the 4th rib cranially from the ilium.

By using the information of Wings the morphology of the epaxial muscles run along the dorsal vertebra and extends to the top of the neural spines would not only thicken the look of the animal, but those with bifurcated neural spines wouldn't have had a bifurcation/indentation during life. *Amargasaurus* is a strange looking sauropod and only the cranial/anterior cervical neural spines would have been free of the epaxial muscles (Though it could argue that like the long neural spines of pleycosaurs which did have a sheath of skin encasing the spines into a fin, *Amargasaurus* long cervical spines also had a sheath of skin because there are several specimens which have broken and healed spines that would have fallen off completely if they lacked a 'fin', therefore the same thing could happen to *Amargasaurus* if it lacked a 'fin'. I originally was against a fin in *Amargasaurus*, but now believe otherwise.)

The dorsal vertebrae are not only different from genus to genus (or family to family) but the number of dorsal vertebrae in dicraeosaurids and diplodocids. *Dicraeosaurus* has the longest back with 12-13 dorsal vertebrae, *Apatosaurus* and *Diplodocus* have a medium length with 10 dorsal vertebrae and *Barosaurus* has the shortest back with 9 dorsal vertebrae. Shunosaurids, dicraeosaurids, diplodocids and rebbachisaurid have the tallest vertebrae, Haplocanthosaurids have neural spines mid height, and camarasaurids and titanosaurids have the shortest.

Because each genus/family has a different skeletal morphology this would indicate a different structural and biological impact on that animal. The epaxial muscles helped to strength and stabilize the

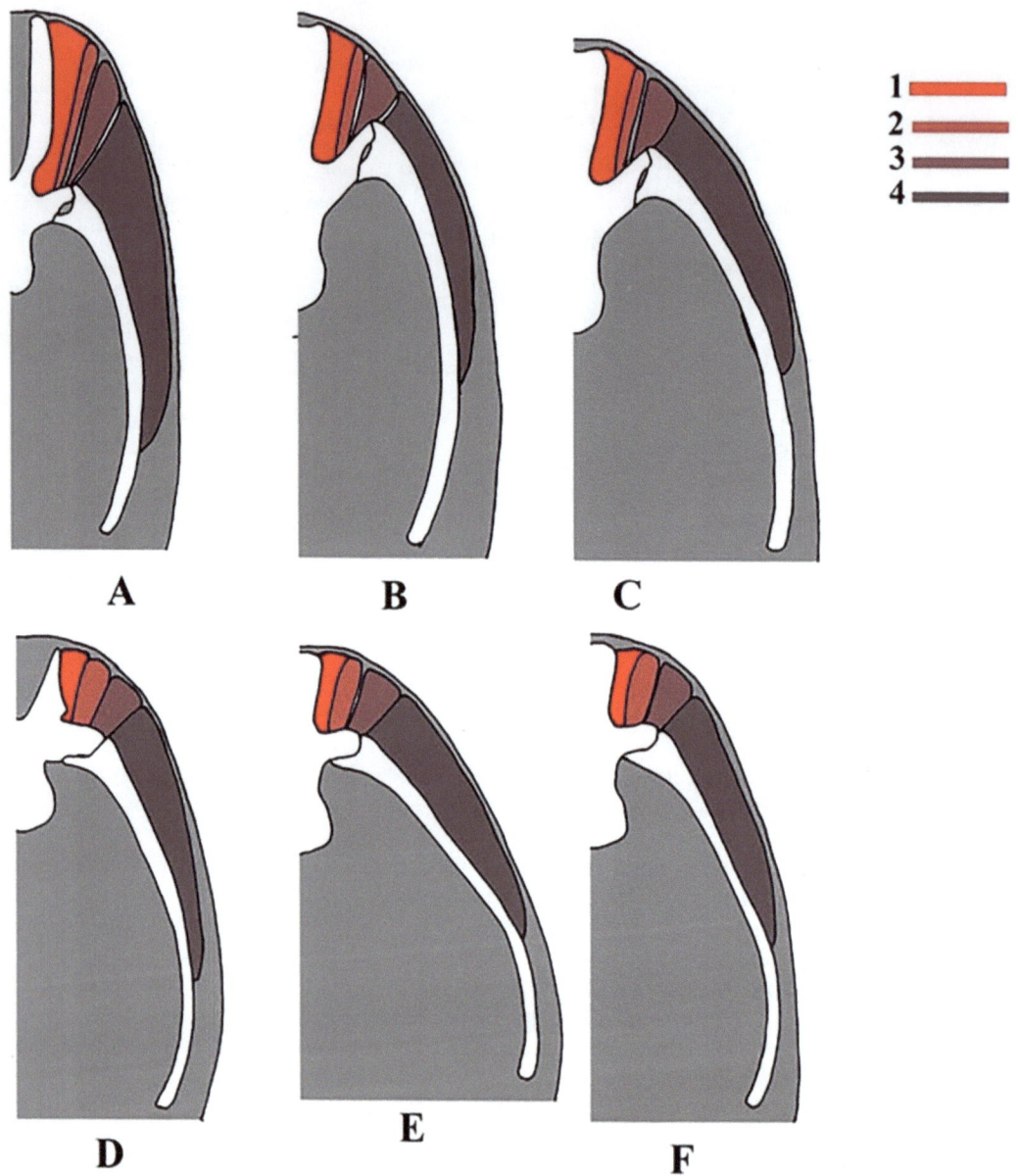

Figure 2) Cross section of the body of sauropods showing the muscles; A-C *Dicraeosaurus*, D *Diplodocus*, E-F *Apatosaurus*; A) *Dicraeosaurus* anterior dorsal vertebra; E) *Dicraeosaurus* middle dorsal; F) *Dicraeosaurus* posterior dorsal; D) *Diplodocus* anterior dorsal vertebra; E) *Apatosaurus* mid dorsal; F) *Apatosaurus* posterior dorsal; color identification of muscles; 1) *m transsp med*; 2) *m. transsp lat*; 3) *m long d*; 4) *m ilcost*.

back. Those with tall neural spines the epaxial muscles help strength the back more than those with short neural spines and would affect how it ate; i.e. browsing (shunosaurids, dicraeosaurids, diplodocids and rebbachisaurid). camarasaurids and titanosaurids have the shortest neural spines and because of the different morphologies of the vertebrae the epaxial muscles acted differently than those of with tall neural spines. *Haplocanthosaurus* neural spines are in between the tall and short neural spine. In Macronaria (*Camarasaurus*) the *m. ilcost* is much thicker than the taller neural spines and would have decreased the caudal direction according the inclination of the transverse processes. Wings doesn't discuss the possibility that some sauropods may have stood on their hind legs or not, but I believe the sauropods with the taller

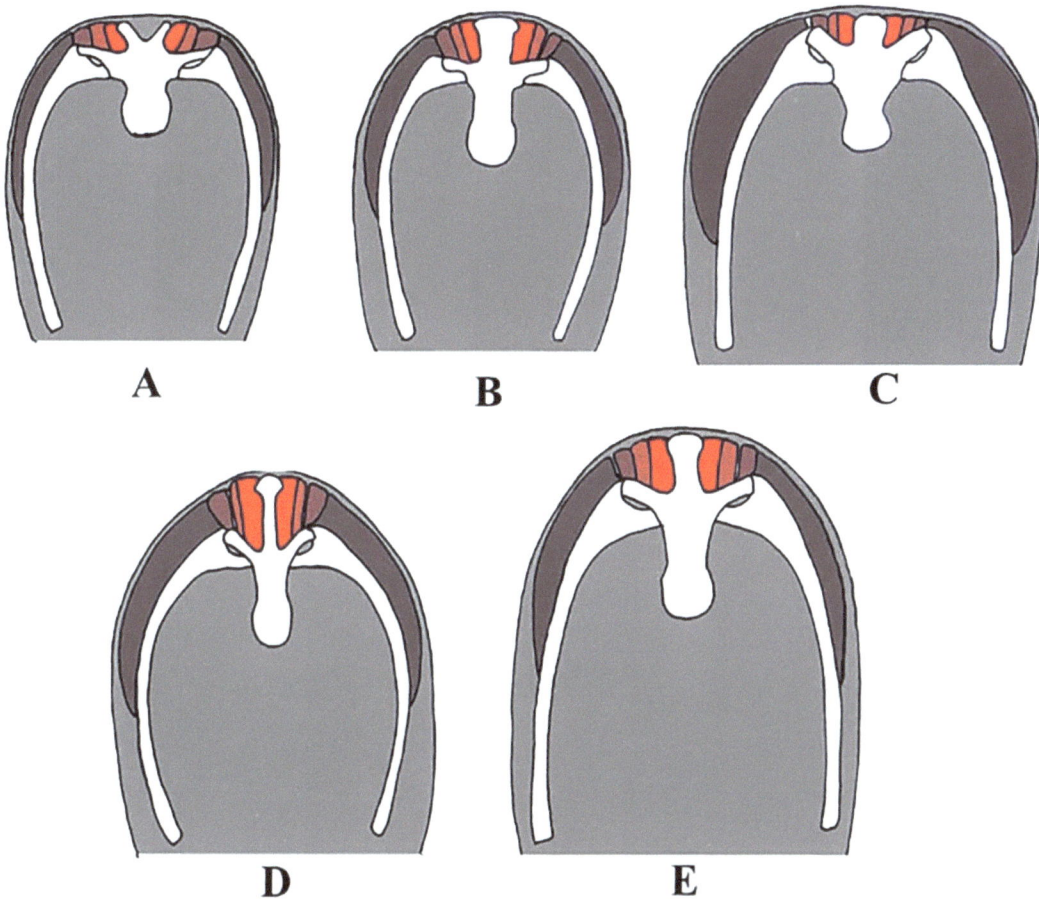

Figure 3) Cross section of the body of sauropods at the mid dorsal region; A-B) *Camarasaurus*; C) *Saltasaurus*; D) *Shunosaurus*; E) *Haplocanthosaurus*. Same muscle color identification as in figure 2.

neural spines and the larger epaxial area would have helped them to be able to rear up on their hind legs. I haven't written about that yet, but I plan on covering that in the next issue.

The top of the neural spines has rugous texture which are similar to those found in birds and crocodiles. This rugosity at the top of the neural spine is where a lattice work of tendons origin is at and inserts into the lower rib articulation (at about a 45-degree angle, give or take) (Figure 4). It isn't known which vertebrae it would insert to because it would vary due to the height of the neural spine. The taller the neural spine the longer the insertion would be. Along the neural spine is a ridge called the spdl (spinodiapophyseal lamina), this is the *m transsp med* origin point. *Dicraeosaurus* has a uniform height in the dorsal neural spines while diplodocids have the tallest neural spines at the posterior dorsals and sacral region. Camarasaurids/titanosaurids have short spines and differ from diplodocids and dicraeosaurids and lacked the stabilization of the tendons that the sauropods with taller neural spines would have. Tendons generally aren't fossilized in sauropods though I do know of one with ossified tendons on the dorsal neural caudal spines region on the *Diplodocus* at the Smithsonian.

What does this all mean? The sauropods with the tall neural spines would have had a trellis-like arrangement of tendons (along the neural spines) and would have not only strengthened the back but also, to some degree stiffen the back (Figure 4). Also, those with tall neural spines the *m. transversospinalis* the center the rotation (intervertebral articulation) would increase. This would not only stabilize the vertebral column but also increase its mechanical advantage (more on this in the next issue). The sauropods with short neural spines (macronarians, titanosaurids) which have a reduction of the *m. longissimus dorsi* and *m. iliocostals*, would have allowed stronger lateral flexibility of the caudal/posterior trunk region. The strong dorsal extension of *m. iliocostalis* would have acted as an efficient lateral flexor of the vertebral column.

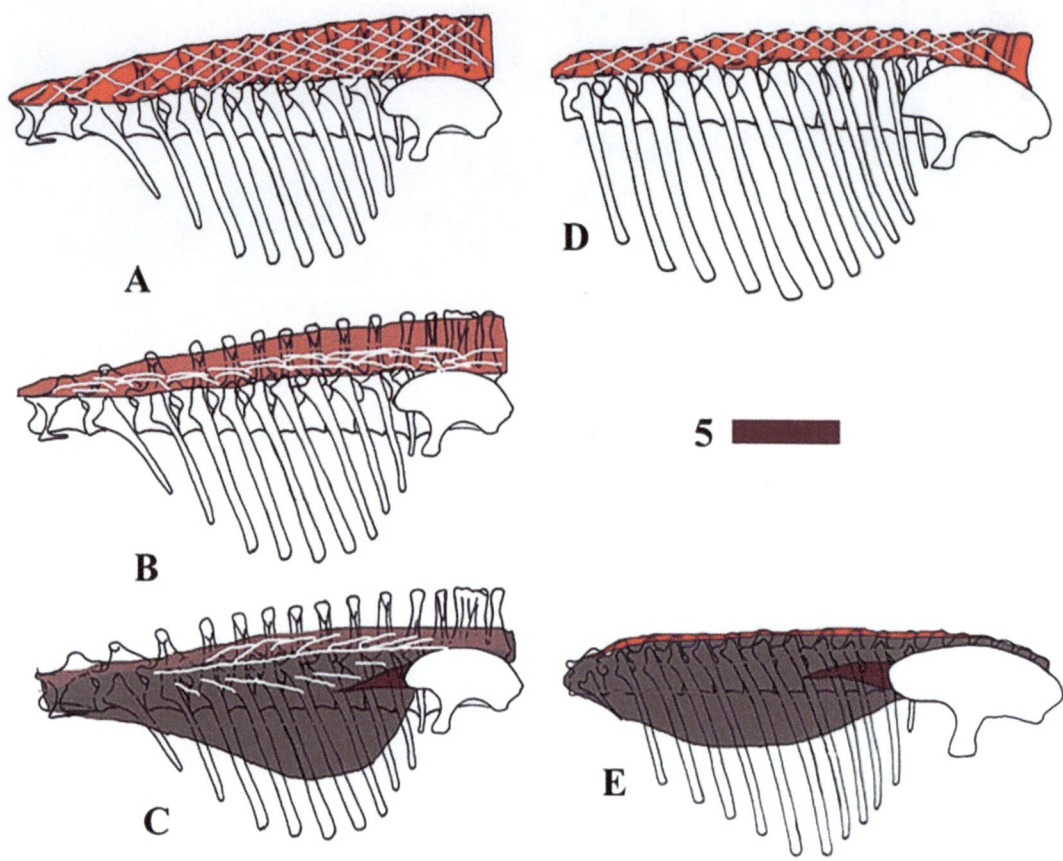

Figure 4) Side view of sauropods showing the muscles and tendons (white stripes); A-C *Diplodocus*; D) *Camarasaurus*; E) *Saltasaurus*. Same muscle color identification as in figure 2 with one more muscle; 5) *ap ilcost*.

This would have worked along with the 'wide-gauged' stance indicating they would have had a different locomotors capability than those with a 'short-gauged' stance.

Traditionally sauropods are illustrated with a well defined neural spines and rib cage, I myself is guilty of this. What happens when the epaxial muscles are added changes the look of sauropods drastically. In shunosaurids, diplodocids, dicreaeosaurids and rebbachiosaurids the body size greatly increases and eliminates the tall neural spines impression. The lean sauropods would then become bulker monstrosity and look nothing like what we are used to (Figure5). It also thickens the rib area/hide and would help deter or help protect the animal from an attack from a theropod.

Bibliography

Schwarz-Wings, D., 2009, Epaxial trunk musculature and ligaments in diplodocids and dicraeosaurids (Dinosauria: Sauropoda) and their postural function during locomotion: In: Journal of Vertebrate Paleontology, v. 29, supplement to n. 3, 69th Annual Meeting, Society of Vertebrate Paleontology and 57th Symposium of Vertebrate Palaeontology and Comparative Anatomy (SPVCA): 178a.

Schwarz-Wings, D., 2009, Reconstruction of the thoracic expaxial musculature of diplodocid and dicraeosaurid sauropods: Journal of Vertebrate Paleontology, v. 29, n. 2, p. 517-534.

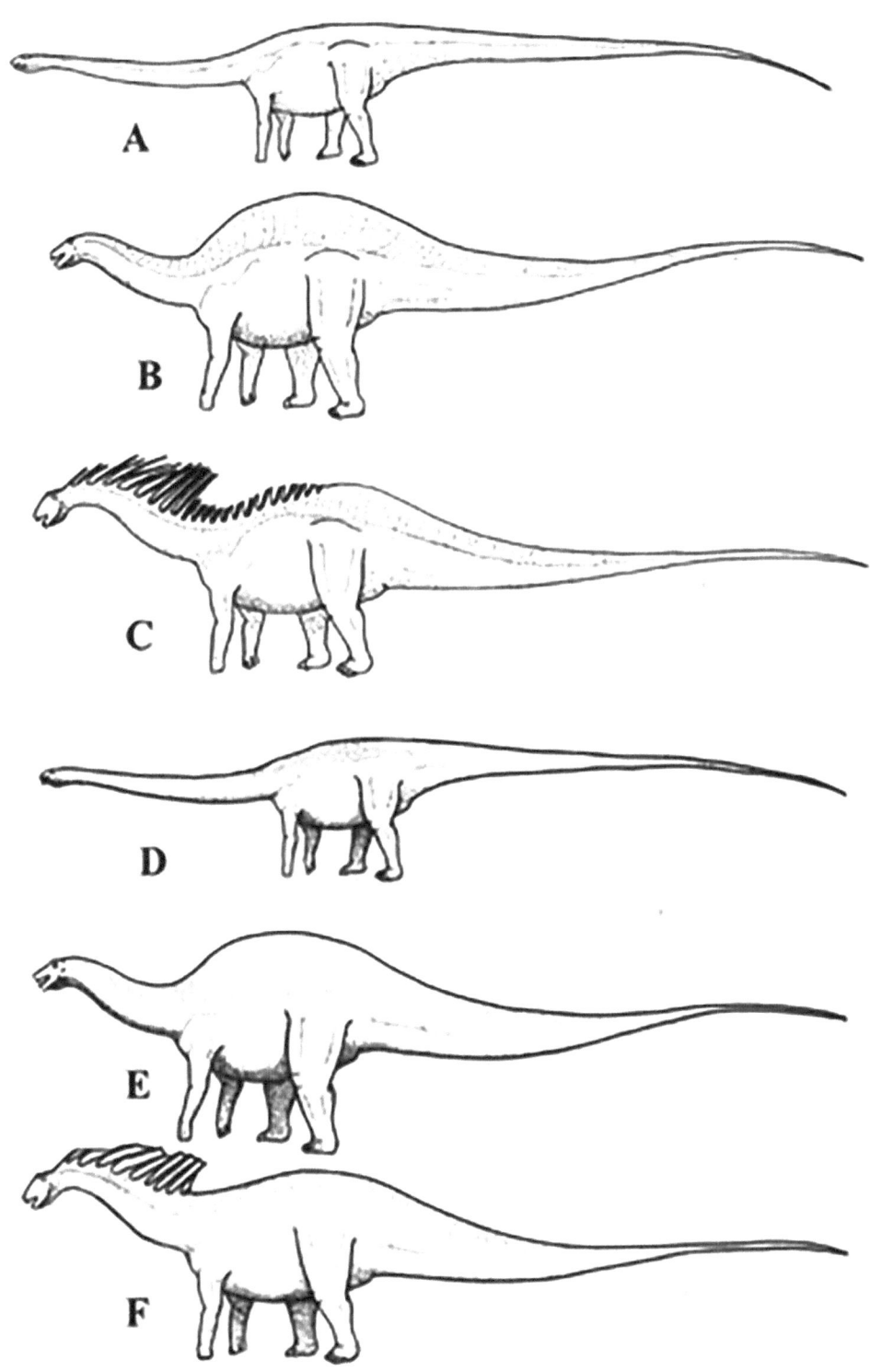

Figure 5) Side view restoration of sauropods, old way (A-C), new way (D-F); A, D); B, E) *Dicraeosaurus*; C, F) *Amargasaurus*.

Ford, T. L., 2010, How to Draw Dinosaurs, Dull or colorfully? What did dinosaurs look like? Prehistoric Times, n. 93, p. 18-19.

Chapter 21

Dull or colorfully? What did dinosaurs look like?

That is an age-old question that goes back to the first discovery of dinosaurs. Dinosaurs were believed to have been big, dimwitted, dull colored reptilian/lizard behemoths. Color doesn't fossilize (only on very rare occasions has color patterns 'fossilized'). So, we just don't know if dinosaurs were dull or colorful. We can infer however, by looking at modern animals what these "behemoths" may have looked like when alive. Only a few mammals are brightly colored, and rest have dull coloration that ranges from brown, red, yellow, grey, black, white or shades in-between them. It is a common misconception that the majority of mammals are color blind. Mammals aren't really color blind, but are more like color 'deficient', i.e. they see color, but not as well as reptiles and birds. Some mammals have color patterns; stripes and spots, that help them blend into their surroundings. A lone tiger walking in the open is easy to spot, its orange fur and black stripes make it stand out, but when it is walking in a dense forest it blends into the foliage and is extremely hard to see. A lone zebra is easy to notice, but when they are in a herd, the strips blend into each other and individual animals are hard to make out. A lot of the color patterns that mammals, and animals in general, have help them blend into their surroundings. Dinosaurs aren't mammals, but I wouldn't argue against comparing similar color patterns. One need only go to a local farm or fair to see the different livestock and their very different color patterns.

Reptiles (Lizards) and birds on the other hand do see colors and many are brightly colored (especially those living in rain forest). Their coloration has different uses; help attract mates, ward off rivals, and also as camouflage. For the most part it is the males that have the brighter colors and the females have duller coloration. Dinosaurs are more closely related to reptiles and birds (actually birds are dinosaurs) and inferring a more colorful world via the use of birds is valid. This would mean the Mesozoic was a bright and colorful world. Even looking at the braincases of dinosaurs the optic lobe is very large and indicates they could see color very well.

But is there any paleontological evidence that proves or disproves colorful dinosaurs? Yes, recently 2 independent studies studying the color of Liaoning theroporods claimed to be able to determine the color of theropods. It has long been established that some dinosaurs were feathered (*Sinosauropteryx*, *Caudipteryx*, *Beipiaosaurus*, *Microraptor*, *Cryptovolans*, etc) and through inference again, we can assume other animals in those families were also feathered, but even that is controversial. The compsognathid *Sinosauropteryx* has 'protofeathers' (but that is still contested by some paleontologist) and *Juravenator* has 'scales'. *Dilong* (actually a referred specimen that may not be *Dilong*) has feathers, but later larger tyrannosaurids didn't (there are specimens with skin impressions that show they had scales).

Both studies (Li et al, 2010, and Zhang, et al, 2010) used scanning electron microscopy (SEM) on different Liaoning theropods and birds. Zhang used *Sinornithosaurus* (IVPP 12811 type) and a referred *Sinosauropteryx* (undescribed, IVPP V14202), a *Confuciusorins* (IVPP V13171), and an isolated feather from Inner Mongolia (IVPP V15388B). Li et al used a referred specimen of *Anchiornis huxleyi* (BMNHC PH828) (Figure 2).

What they were looking for was melanosomes (which are lysosome-related organelles that determine color in modern birds) which are found in the feathers. Phaeomelanin is reddish-brown to yellow (having globular structures), and eumelanin (having rod like structure) are black-grey pigments. The authors found the melanosomes deeply imbedded in the feather structure which indicates they aren't bacteria (on the feathers barbules) (Figure 1). If it was bacteria the structures would be coating the feathers and surrounding matrix not within the feathers themselves. The melanosomes were also tightly packed and layered in the feathers that look like modern feathers. The SEM's also indicate that what they are looking at is actual feathers and not bacteria or collagen fibers.

What colors did they find? And does this have anything to do with the evolution of feathers? Zhang et al argue that the filaments on the referred *Sinosauropteryx* aren't collegian fibers and are 'protofeathers' and are probably evolutionary precursors to feathers. I don't understand the line of logic since the Liaoning biota is millions of years after *Archaeopteryx* and in my opinion can't be part of the evolution of feathers. But *Anchiornis* on the other hand is Oxfordian in age, which is earlier than *Archaeopteryx*. The new specimen of *Anchiornis* indicates it had feathers on its wings, hind legs (4 winged

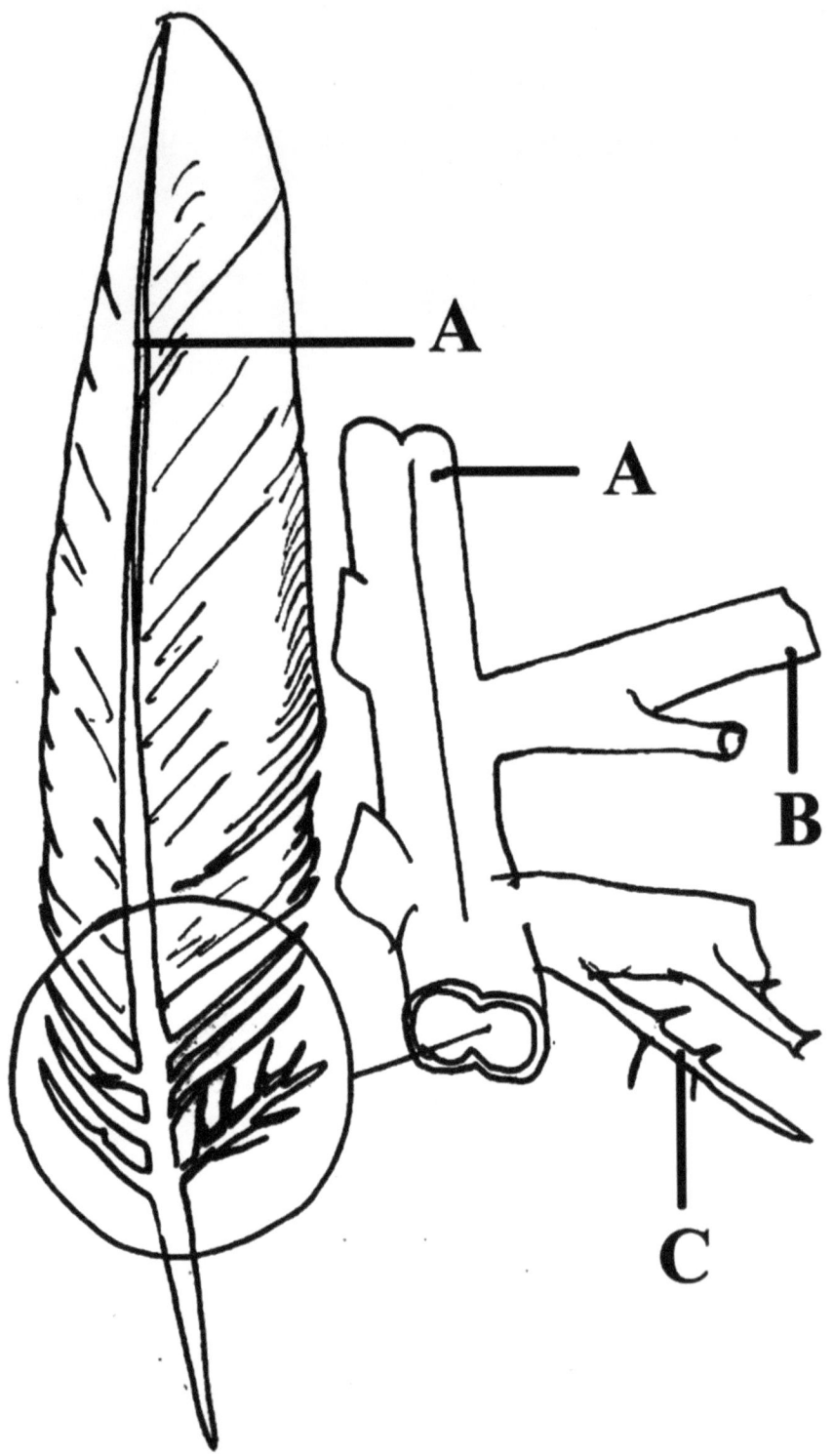

Figure 1) Feather diagram showing the placement of barbules. A) Shaft; B) Barb; C) Barbules (after Proctor & Lynch, 1993).

theropod) and all over its body. Before these theropods were found they were thought to be 'typical' theropods with scales and basically an unornamented body (Figure 3). *Anchiornis* shows that the head had extensive plumage crown which formed a crest. The crest varies from about 1.8-1.4 centimeters, the body

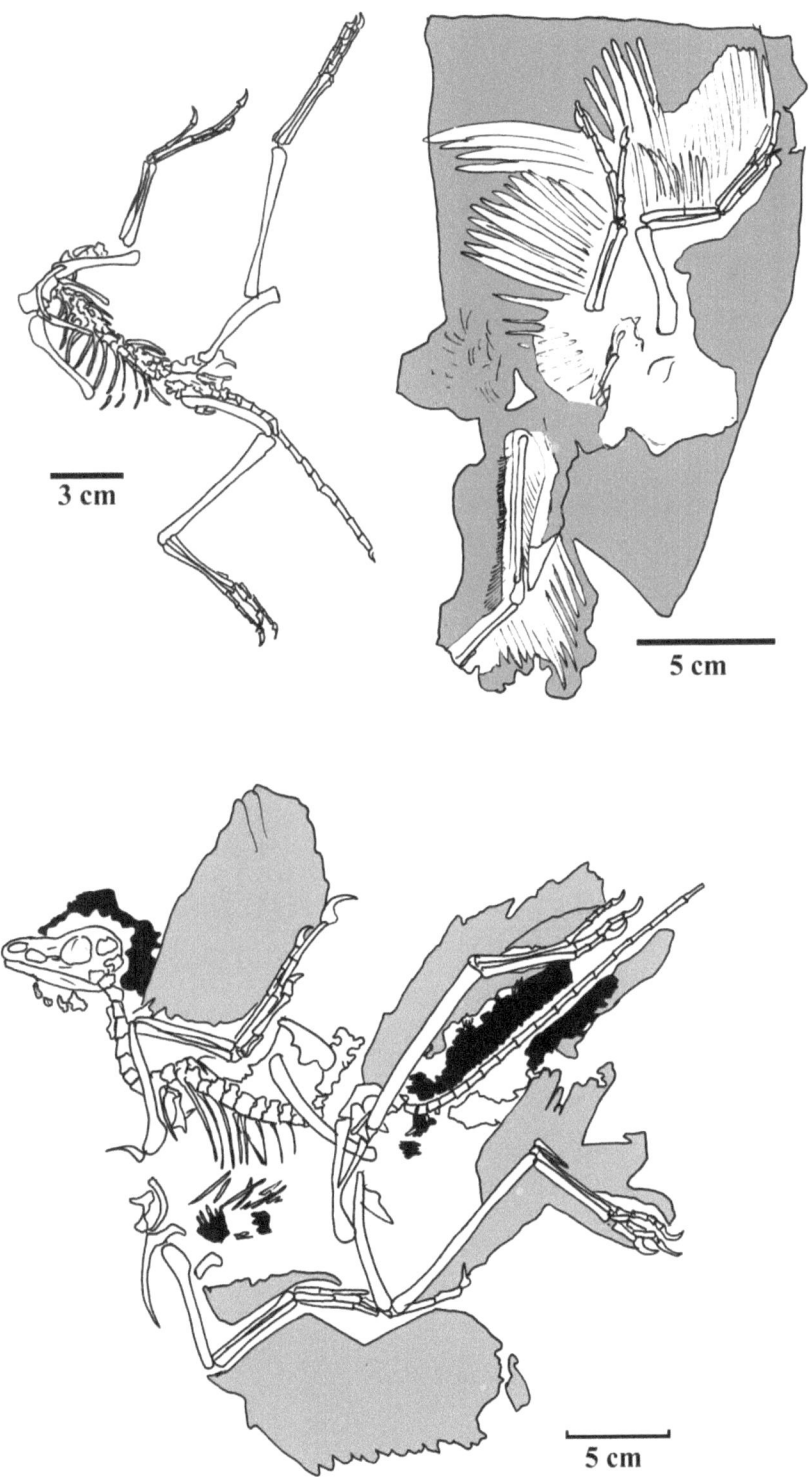

Figure 2) Skeletons of *Anchiornis huxleyi*; A) Type, IVPP V14378; B) Referred specimen BMHNC PH828; C) Referred specimen LPM-B00169.

feathers were 1.5 centimeters long. The wing feathers are interesting in that the feathers themselves are narrower than modern flying birds and couldn't have sustained flight. There are more than 12 primaries, 22-23 secondary and the 'wings' are like modern birds in their morphology.

As far as coloring *Sinosauropteryx* (which they were only able to study the tail) had dark-coloured stripes and a possible filamentous crest along the back in chestnut to rufous (reddish-brown) tones. *Sinornithosaurus* filaments have locally dominate eumelanosomes or phaeomelanosomes which indicates they had signification different color tones. To me this means they were very colorful. Zhang et al, didn't go into what the feathers were used for, communication, camouflage or thermoregulation. I have no problem with the first two since the banding on the tail would help them to blend into their surroundings, but I have a problem with thermoregulation with the world being a hot house in the Mesozoic, why would you need to keep warm if the world was already warm? But I digress…

Anchiornis is very interesting in its coloring. Its body plumage was darkly colored in black/grey. The head was grey with rufous and black and it had an elongate crest with grey feathers on the sides and front while the back may have been a reddish brown. The wing coverts had a grey margin with a dark epaulet which contrasted strongly with black/grey-spangled lighter primaries, secondaires and greater coverts on the forelimb plumage. The spangles on the outer-most primaries were black. The greater coverts on the upper wing were spangled with grey or black, the secondary and primary rows appeared as conspicuous dots. The contour feathers on the legs (the shanks) were grey, and black on the feet. The hind limb elongate feathers were white at their bases with broad black distal spangles. Unfortunately, the tail is unknown in this specimen. Because the wing feathers weren't used for flight, the coloring and feathers could function as interspecific threat and defensive postures like in many modern birds. Finding feathered theropods older than *Archaeopteryx* would play a major role in the evolution of not only feathers but also flight. More research needs to be done to help determine this.

One more recent article by Hone, et al studied the wing feathers in the type *Microraptor gui*. They used ultraviolet light and to put it in a nut shell showed that the wing feathers (fore and hind) are indeed real feathers and in their natural position.

So what does all this mean? Simple, the old view of dromeosaurids, troodontids, oviraptorids, etc of scaly, 'thin' theropods should be replaced with feathery 'thicker' theropods and dinosaurs in general should be illustrated not with dull colors, but as brighter, more colorful animals. Luis Rey championed this many years ago with his colorful dinosaurs/animals of the Mesozoic. His beautiful brightly colored paintings of the Mesozoic world pioneered a colorful world in which others kept dull. I would recommend anyone to follow in his footsteps and show the Mesozoic in a more colorful light.

And to quote Dr. Thom Holz from the DML (Dinosaur Mailing List): We are truly living in the golden age of dinosaur discoveries!

Bibliography

Li, Q., Gao, K.-Q., Meng, Clarke, Shawkey, D'Alba, L., Pei, R., Ellison, M., Norell, M. A., and Vinther, J., 2012, Reconstruction of *Microraptor* and the evolution of iridescent Plumage: Science, v. 335, p. 1215-1219.

Proctor, N. S., and Lynch, P. J., 1993, *Manual of ornithology, avian structure & function. Yale University Press, 340pp.*

Zhang, F., Kearns, S. L., Orr, P. J., Benton, M. J., Zhou, Z., Johnson, D., Xu, X., and Wang, X., 2010, Fossilized melanosomes and the colour of Cretacoeus dinosaurs and birds: Nature, published online, 4pp.

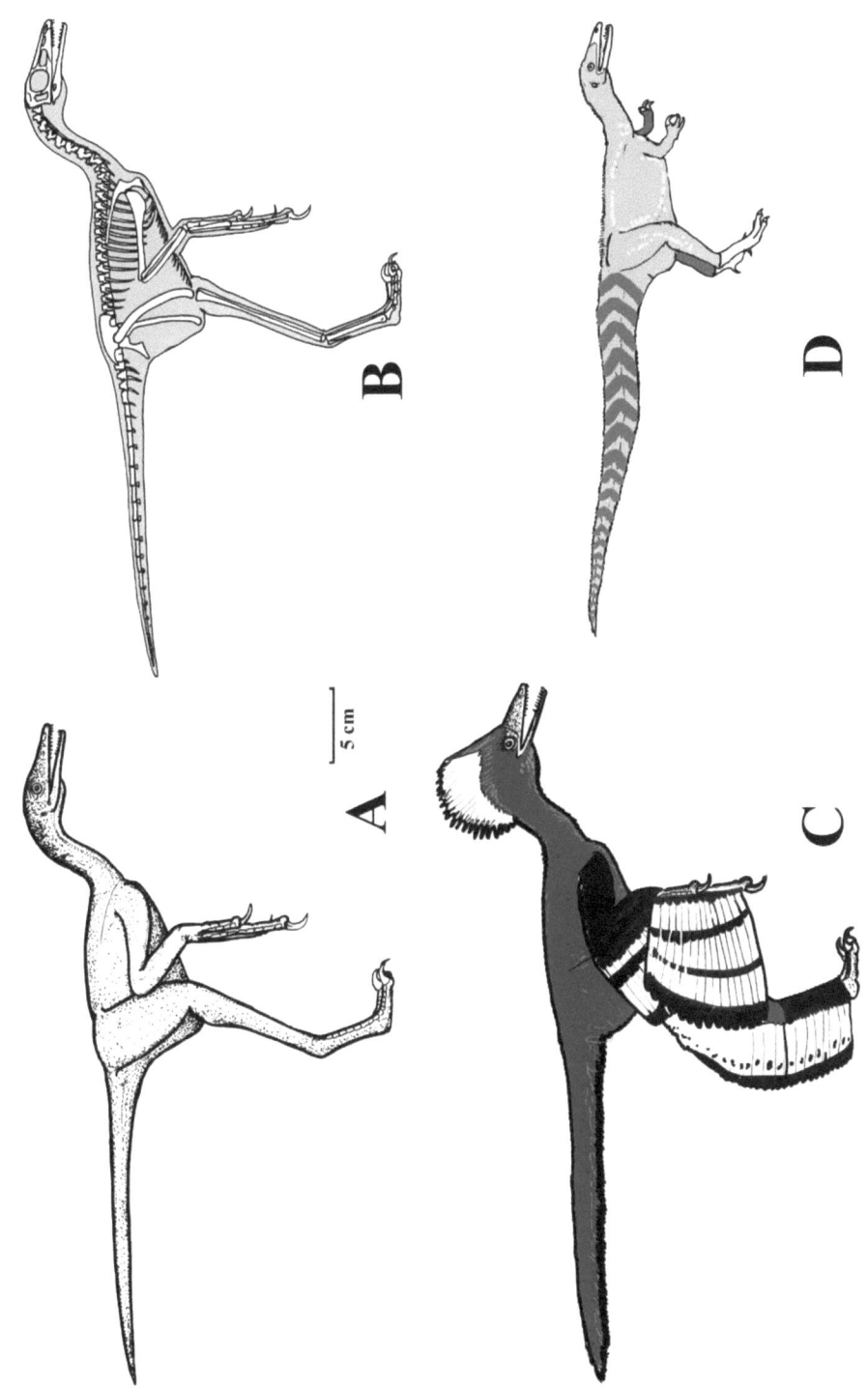

Figure 3) Reconstructions of *Anchiornis* and *Sinosauropteryx*; A) Life restoration of *Anchiornis* (LPM-B00169) in the old version of scaly dinosaurs; B) Skeleton of referred specimen LPM-B00169; C) Life restoration of *Anchiornis* (LPM-B00169) using recent studies showing the color pattern based on BMHNC PH828; D) *Sinosauropteryx* showing the color pattern on the tail.

Ford, T. L., 2010, How to Draw Dinosaurs, Aquatic Pittacosaurs (finally) Part one. Prehistoric Times, n. 94, p. 18-19.

Chapter 22

Aquatic Pittacosaurs (finally) Part one

Now that the Ceratopsian volume has been published (New Perspectives on Horned Dinosaurs: The Royal Tyrrell Museum Ceratopsian Symposium (Life of the Past), Indiana University Press). I can now write my article on aquatic psittacosaurs. The idea for the article started when I was talking to Larry Martin (my co author) at the Tucson Rock Show in 2007. He was there for a talk he gave for the AAPS the night before (I missed it). He told me years ago he saw a small Chinese Salamander being sold with a wide tail. He thought it was misidentified and was actually a juvenile *Psittacosaurus*. Unfortunately, he didn't buy it, though I was able to track down who sold it and whom he sold it too, but the dealer didn't remember who he sold it too. This got me thinking and I thought we could do a joint paper on the possibility of aquatic psittacosaurs and present it at the 2007 Ceratopsian symposium being held later that year at the Royal Tyrrell museum of Paleontology. He agreed to co-author the article, but he told me I was to be the senior author. I was a little leery about being the senior author but agreed to write the article. It turns out we aren't the first people to theorize psittacosaurs were aquatic; Rozhdestvensky, Suslov, Averianov and Currie beat us to it, individually. (I had to ask Currie to make sure I didn't read it wrong, but he said he did write it.)

Rozhdestvensky (1955) states that the manus phalanges and unguals are flat and may have had webbing between the fingers. He also thought that the sclerotic rings indicated an aquatic lifestyle. The anatomy of the hand would also have made it easy for *Psittacosaurus* to swim.

Suslov, (1983), believed that psittacosaurs either lived near or in water due to occurrence of many specimens being found in lacustrine deposits of Khamryn-Us locality in Mongolia.

Currie (1997, pers. comm..) believes that the reason why psittacosaurs had gastroliths was for ballast and not for eating. The teeth had a strong self-sharpening cutting edge and would have been good for cutting, and the gastroliths would not have been of much additional help in digesting food.

Averianov et al. (2006) followed the interpretation of this group of authors, also the dorsally flattened phalanges were good for walking on soft substrate.

And we can add Ford & Martin, (2010), what the above wrote, as well as dorsally placed nares and orbits, sprawling hind limbs, and swimming stroke range of motion of the fore limb.

Interestingly more psittacosaur specimens have been found in lacustrine (Lake) deposits than terrestrial (i.e. dune deposits). Many have been found in three large Early Cretaceous lakes; two in northern China: Qingyang Lake on the modern Ordos Plateau, a large lake in the Junggar Basin (Averianov et al, 2006, Chen 1987) and the third in Khamryn-Us, Mongolia (Suslov 1983).

In writing up the article and doing literature research I noticed that psittacosaurs are found in two different positions, which I call life positions; resting and sprawling (Figure 1). The first and type species of *Psittacosaurus*, *P. mongoliensis*, was found with its arms at its sides, slightly bent with the palms up, the hind legs folded (flexed and abducted). The knee is in a hyperflexed knee (Sereno and others incorrectly called this hyperextended knee, and this was pointed out to me by my peer reviewer Ralph Molnar) (also called a sharp Z' by Coombs 1982) as if it was squatting down. Coombs believes this posture is post-mortem and not a true resting behavior. When the animal died, it settled down on its belly with the hind limbs folded into a sharp 'Z', the knee projecting forward, the ankle backward, femur atop of the tibia/fibula, and pes firmly pressed flat against the ground. Faux and Padian determined that the position of *P. mongoliensis* is not consistent with opisthotonic posture and is a natural position. A referred specimen of *Psittacosaurus xinjiangensis* though not as complete, also has its hind legs folded and was interpreted as being in a resting position (As per Brinkman et al). In fact, any dinosaur found in this position can be considered found in a resting position: I've seen this in theropods, ornithopods and yes even in Sauropods. The Wealden small ornithopod *Stenopelix valdensis* was also found in a resting position and I believe *Stenopelix* is a psittacosaurid (which was taken out of the manuscript, so this will be something not in the article). If specimens found in a resting position is a natural life position what about sprawling?

Sprawling is another position that psittacosaurs are often found in. The hind legs are horizontal to the body, with the lower legs perpendicular to the body. The fore leg can also be in the same position. The second psittacosaur specimen described, *Protiguanodon mongoliensis* had sprawling hind legs with its arms

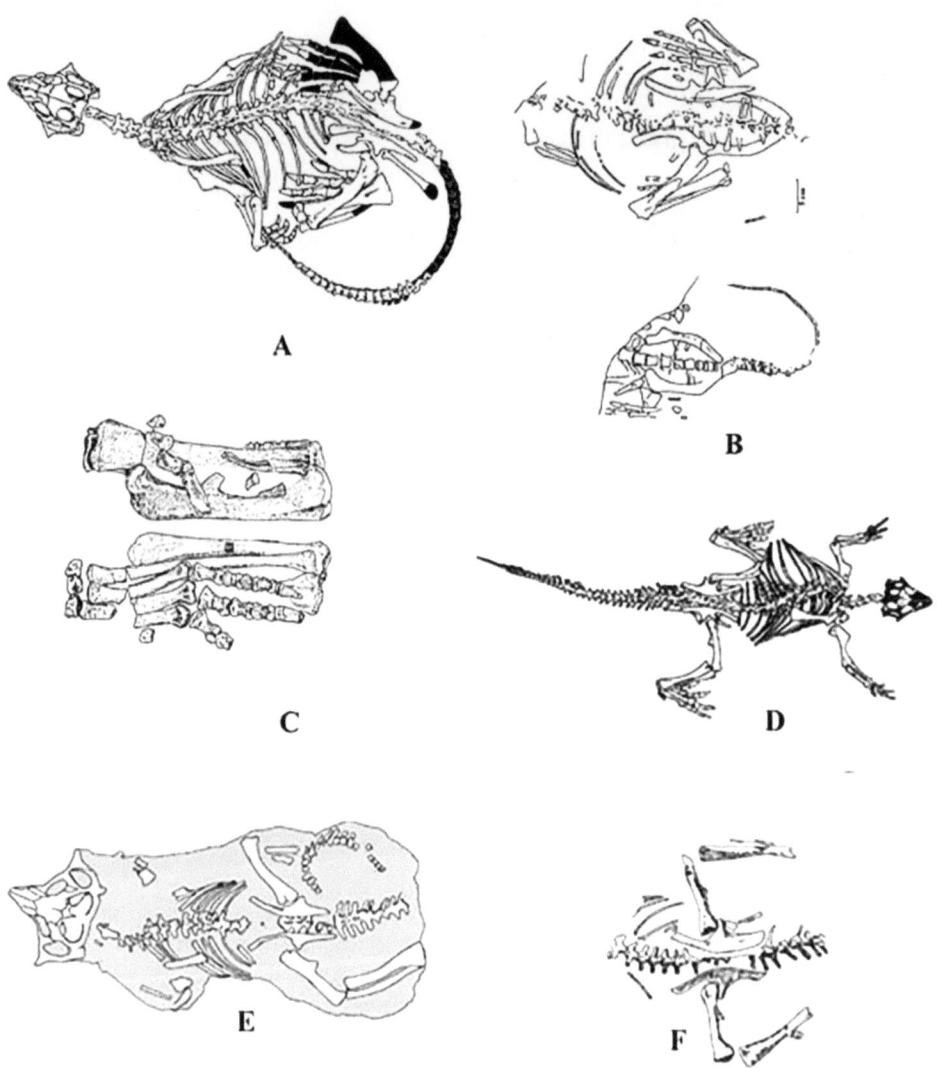

Figure 1) Resting (A-C), and sprawling (D, E) psittacosaurs. A) *Pisttacosaurus mongoliensis*. Dorsal view; B) *Stenopelix valdensis*. Dorsal and ventral view; C) *Psittacosaurus xinjiangensis*, right hind limb, in dorsal and ventral view: D) *Psittacosaruus mongoliensis = Protiguanodon mongoliensis*, dorsal view; E, F), *Psittacosaurus sinensis*, dorsal view.

outstretched (Senter 2007 believes the arm is not in a natural state and is dislocated from the glenoid). Several Psittacosaurs have been found in the sprawling position from juveniles to adults.

Psittacosaurs are believed to have been bipedal animals, with erect limbs, but could they sprawl? Is there any evidence they could? *Iguanodon* and for that matter ornithopods in generally have their femoral heads at a right angle to the femur, in psittacosaurs is at a 30° angle and the top of the femoral head shows evidence of a large cartilaginous cap (Figure 2). This gives the femur a larger range of movement, vertical for bipedality and lateral/horizontal (or sprawling) for swimming. This would enable them to kick their feet like swimming mammals and reptiles. The femoral head of *Stenopelix valdensis* is also at a 30° which support its psittacosaurid affinities. It has been pointed out to me that ceratopsians in general have a cartilage cap on the femur head, but this isd due to the fact that the ilium has a 'flange' that hangs over the accitabulum. Psittacosaurs lack this flange.

The feet (both front and hind) are in unique in psittacosaurs. The pes unguals and phalanges are dorso-ventrally compressed. The length of the toes are misleading, in that digits III and IV look to be the

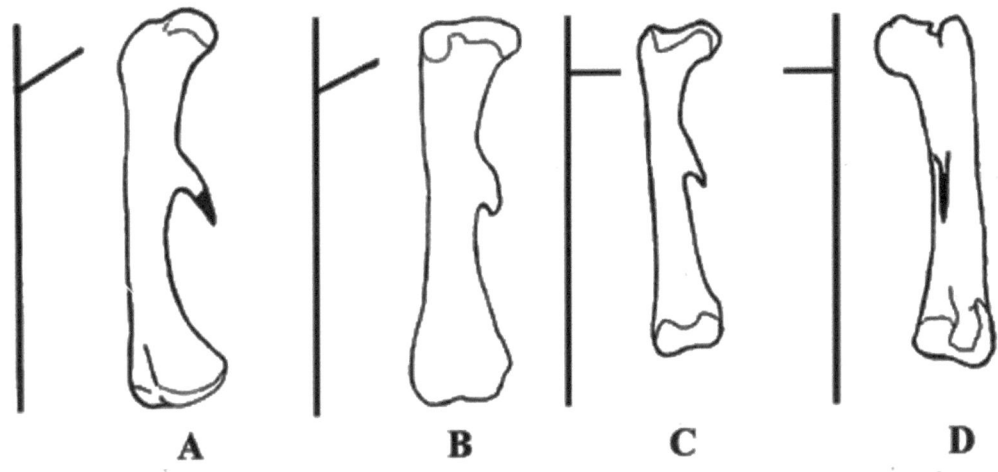

Figure 2) Ornithopod femurs. A) *Psittacosaurus xinjiangensis*; B) *Psittacosaurus sibiricus*; C) *Hypsilophodon foxii;* D) *Iguanodon bernissartensis*.

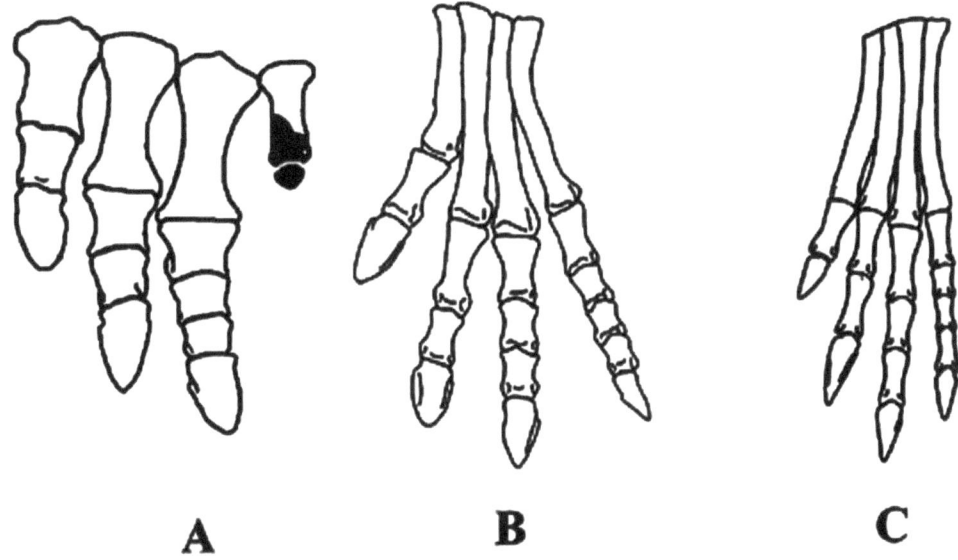

Figure 3) Manus and pes of Psittacosaurs. A) Manus of *Psittacosaurus sibiricus*; B) Pes of *Psittacosaurus neimongoliensis*; B) Pes of *Stenopelix valdensis*.

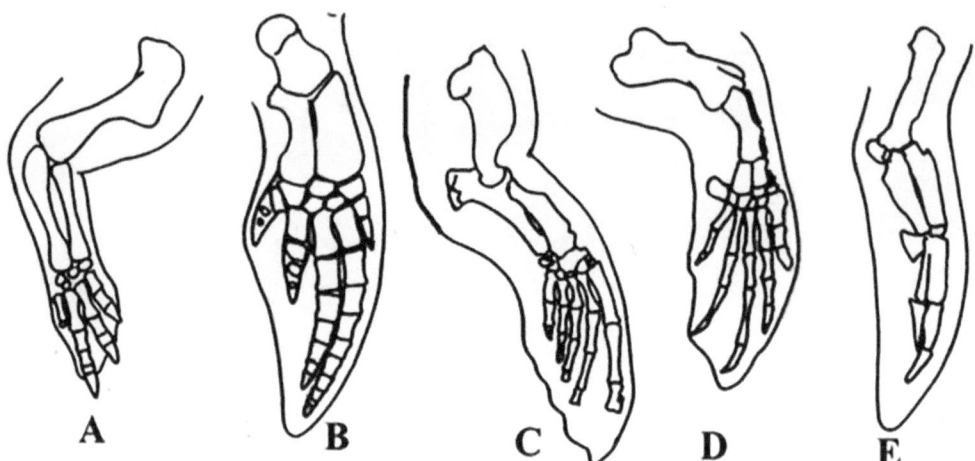

Figure 4) Front legs of swimming animals. A) *Psittacosaurus*; B) Dolphin (*Lagenorhynchus*); C) Sea Lion (*Zalophus*): D) Sea Turtle (*Cheloni*); E) Penguin, (*Spheniscus*).

same length. This is because the metatarsals are of different length (MT III is longer than MT IV), but the phalanges and unguals are the same length. The only other dinosaurs that has this are protoceratopsids and *Stenopelix* (another reason why I believe it's a psittacosaurid) and may be a diagnostic feature in basal margenocephalians. It was proposed that the hind toes also had webbing between them by Rozhdestvensky.

The manus is very strange. The longest finger is digit III and not digit II as seen in all other ornithopods. Unfortunately, the manus of *Stenopelix* isn't known, but if it was the same as *Psittacosaurus* than that would be a slam dunk psittacosaurid. We proposed that the forefoot could have been webbed and may have looked like a paddle. Sereno and Chinnery believed that they used their front feet in grabbing and pulling foot toward its mouth, but Senter believed otherwise and will be explained in the next part.

The manus shows some similarity to swimming animals; penguins, seals and dolphins, although compared to those animals the longest digit is reversed. The leading edge of eh first digit (seals) or second digit (dolphins) is the longest and strongest and is used heavily in swimming. If the and was webbed as Rozhdestvensky suggested, the manus would not have been as strong a swimming device as in other marine mammals/lizards. The metacarpals are parallel and are close to each other. Another possibility is that the hand may have been held together by thickened skin like that of a sea turtle (Figure 4).

Bibliography

Brinkman, D. B., Eberth, D. A., Ryan, M. J., and Chen, P.-J., 2001. The occurrence of *Psittacosaurus xinjiangensis* Sereno and Chow, 1988 in the Urho area, Junggar Basin, Xinjiang, People's Republic of China: In The Sino-Canadian Dinosaur Project 3, Canadian Journal of Earth Sciences, 38, pp. 1781-1786.

Currie, P. J., 1997, Chinese Dinosaurs: International Dinosaur Symposium from Uhangri Dinosaur center and Theme Park in Korea, edited by Yang, S.-Y., Huh, M., Lee, Y.-N., and Lockley, M. G., The Paleontological Society of Korea, Special Publication n. 2, p. 93-101.

Galton, P. M., 1974. The Ornithischian Dinosaur *Hypsilophodon* from the Wealden of the Isle of Wight. Bulletin of the British Museum (Natural History), Geological Series, 25, 1: 3-152.

Norman, D. B., 1980. On the ornithischian dinosaur *Iguanodon bernissartensis* of Bernissart (Belgium). Institut Royal des Sciences Naturelles de Belgique, Memore, 178. 1-103.

Osborn, H. F., 1924, *Psittacosaurus* and *Protiguandon*: Two Lower Cretaceous Iguanodonts from Mongolia. American Museum Novitates, 127: 1-15.

Rozhdestvensky, A. K., 1955. New data on *Psittacosaurus*-Cretaceous ornithopods. In Questions on the geology of Asia, 2: 783-788.

Schmidt, H., 1969. Stenopelix valdensis H. v. Meyer, der kleine Dinosaurier des norddeutschen Wealden. Paläontologische Zeitschrift, 43: 194-198.

Senter, P., 1970. Analysis of forelimb function in basal ceratopsians. Journal of Zoology, published online: 10pp.

Sereno, P. C., and Chao S., 1988. *Psittacosaurus xinjiangensis* (Ornithischia: Ceratopsia), a new psittacosaur from the Lower Cretaceous of northwestern China. Journal of Vertebrate Paleontology, 8, 4: 353-365.

Sues, H.-D., and Galton, P. M., 1982. The systematic position of *Stenopelix valdensis* (Reptilia: Ornithischia) from the Wealden of North-Western Germany. Palaeontographica, Abt. A, 178: 183-190.

Suslov, J. V., 1983. The locality of *Psittacosaurus* in Chamrin-us (East Gobi, MPR). Transactions of the Joint Soviet Mongolian Paleontological Expedition, 24: 118-121.

Young, C.-C., 1958. The Dinosaurian Remains of Laiyang, Shantung. Palaeontologia Sincia, 142, C, 16: 1-138.

Ford, T. L., 2010, How to Draw Dinosaurs, Aquatic Pittacosaurs (finally) Part two. Prehistoric Times, n. 95, p. 18-19.

Chapter 23

Aquatic Pittacosaurs (finally) Part two

And I continue…

Psittacosaurus sp (SMF R 4970) is the specimen with the 'bristle-like' structures on the anterior dorsal portion of the tail (Figure 1). They are approximately 100 of these long thin structures. It also has skin impressions over almost the complete specimen. We theorized that there is a possibility that the bristle along the anterior part of the tail may have been encased in flesh like in some aquatic salamanders. Even though SMF R 4970 has a body outline it has none around the bristles, which may or may not indicate it did or didn't have a flesh tail. This could be due to the preservation or preparation. We came to this conclusion from Larry Martins identification of the small salamander being sold at Tucson may have actually been a baby psittacosaur (though he didn't buy it and was working off his memory).

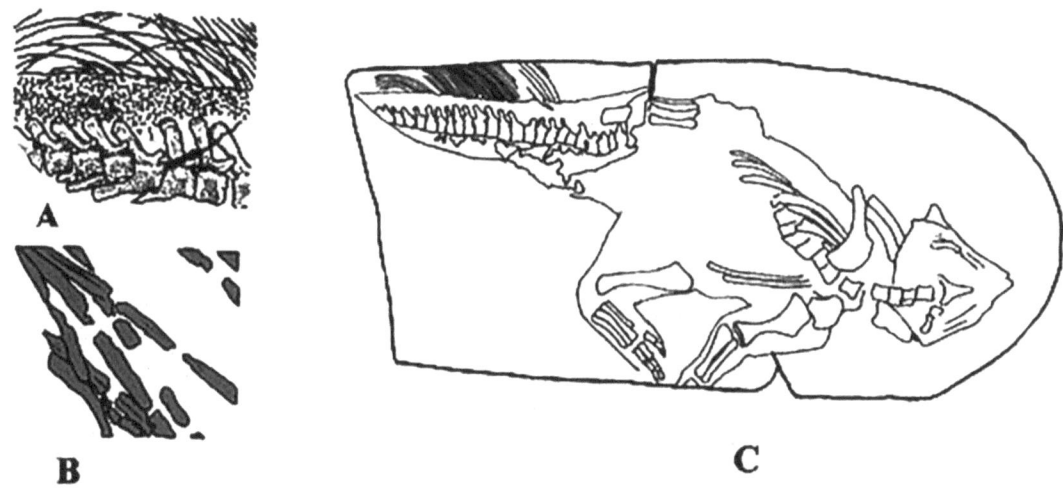

Figure 1), Bristle tail of *Psittacosaurus*.

The manus shows some similarity to swimming animals; penguins, seals and dolphins, though compared to those animals the longest digit is reversed. The leading edge of the first digit (seals) or second digit (dolphins) is the longest and more robust. If the hand was webbed as Rozhdestvensky suggested, the manus would have been a strong swimming device as in other marine mammals/lizards. In dolphins the front fins are used for steering and in seals the front fins are used as major propulsion. In *Psittacosaurus* the metacarpals are parallel to one other and the phalanges are close to each other. I don't think they could be spread out and webbed. Another possibility is that the hand may have been held together by thicken skin like that of a sea turtle.

The clincher for me was a paper written by Senter on the forelimb function in basal ceratopsians; protoceratopsids and psittacosaurids. He found the shoulder range of *Psittacosaurus* consists of parasagittal movements that have a wide arc, but protraction was limited to not much more than a horizontal position (Figure 2). The transverse movement of the humerus allows it to be raised to the subhorizontial position. When the humerus is parallel to the body, 21° from horizontal to the body and the ulna/radius is at 55° (the full extent the ulna/radius can move) the humerus can be moved forward from the 21° to horizontal. The ulna/radius can be moved 93° and when the humerus is brought back it can be raised 64° from the original position. This is an unusual motion for a terrestrial animal, as they could not reach forward or bring food to their mouth or dig as has been suggested by Sereno and Chinnery. Such a limited range of fore limb movement does not support gathering food to the mouth or scratching body parts or used for walking. Senter's explanation for this movement is for clutching objects to the body. We believe the

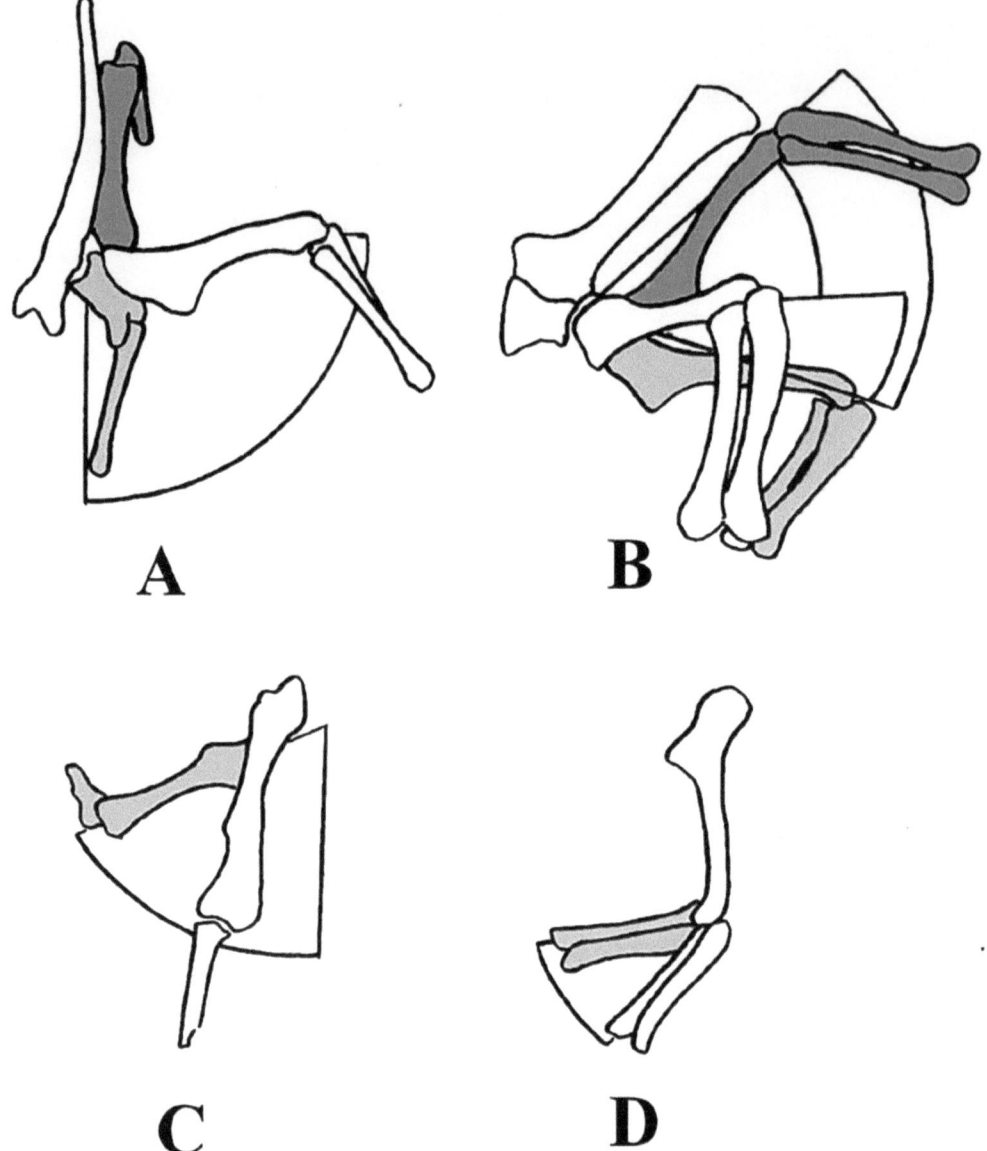

Figure 2). Range of movement of the pectoral girdle of *Psittacosaurus neimongoliensis*. A) cranial view; B) lateral view; C) dorsal view; D) lateral view of forelimb. All after Senter 2007.

explanation for the range of movement was for swimming and his work shows an ideal range of a swimming stroke.

Psittacosaurs are the most abundant dinosaur known, more so then hadrosaurids and ceratopsians. Dodson estimated 120 *Psittacosaurus* specimens, but this was before the wave of fossils from Liaoning. Lu, et al. (2007) state more than 200 specimens have been found in western Liaoning alone. Recently Dr. Horner has had expeditions to Mongolia in search for *Psittacosaurus* for on going research on dinosaur ontogeny studies and had found 67 skeletons and was expected to collect another 100 or so specimens. They have been found in Valanginian to Albian Early Cretaceous deposits and have been found in Mongolia, China, Siberia, and Thailand. There are 13 different species named (The most abundant in any dinosaur). What is interesting is that they also have different morphologies, some had a large head to body

Figure 3). Psittacosaur skeletons in lateral view all to the same size. A) *Psittacosaurus mongoliensis*; B) *Psittacosaurus mongoliensis = Protiguanodon mongoliensis*; C) *Psittacosaurus sinensis*; D) *Psittacosaurus neimongoliensis;* E) *Psittacosaurus* sp (SMF R 4970); F) *Psittacosaurus xinjiangensis*; G) *Psittacosaurus sinensis = Psittacosaurus youngi*; H) *Psittacosaurus major*); I) *Psittacosaurus sibiricus*; J) *Stenopelix valdensis*.

Figure 3a). Psittacosaur skeletons in lateral view all to the same scale. A) *Psittacosaurus mongoliensis*; B) *Psittacosaurus mongoliensis = Protiguanodon mongoliensis*; C) *Psittacosaurus sinensis*; D) *Psittacosaurus neimongoliensis;* E) *Psittacosaurus* sp (SMF R 4970); F) *Psittacosaurus xinjiangensis*; G) *Psittacosaurus sinensis = Psittacosaurus youngi*; H) *Psittacosaurus major*); and I) *Psittacosaurus sibiricus*.

size, some had front and hind limbs nearly equal to each other and others had longer hind limbs. These different proportions may indicate different life styles (Figure 3).

A few other possible aquatic features are dorsally placed naris and orbits and gastroliths. Currie believed the teeth were sharp enough to cut up food and they didn't need the gastroliths for digestion and may have used them for ballasts. The orbits are dorsally placed on the skull as is the naris. Aquatic animals have eyes dorsally placed which allows the animal to keep the majority of its body below the surface of the water. Capybara also have dorsally place eyes and naris and we compare the skull of Capybara to *Psittacosaurus*. It is interesting to note that some *Psittacosaurus* (*P. sinensis*) orbits were foreword facing and had more of a stereoscopic vision than *Tyrannosaurus rex* (Figure 4). It some ways the skull of *Psittacosaurus* reminds me of dicynodonts. *Psittacosaurus* may have used its tail in a similar manner as crocodilians. Crocodilians swim in an undulatory or anguilliform and they would have swum in the same way. The sprawled legs which they used for kicking and the forelimb indicate it had a strong swimming stroke as indicated by Senter's research.

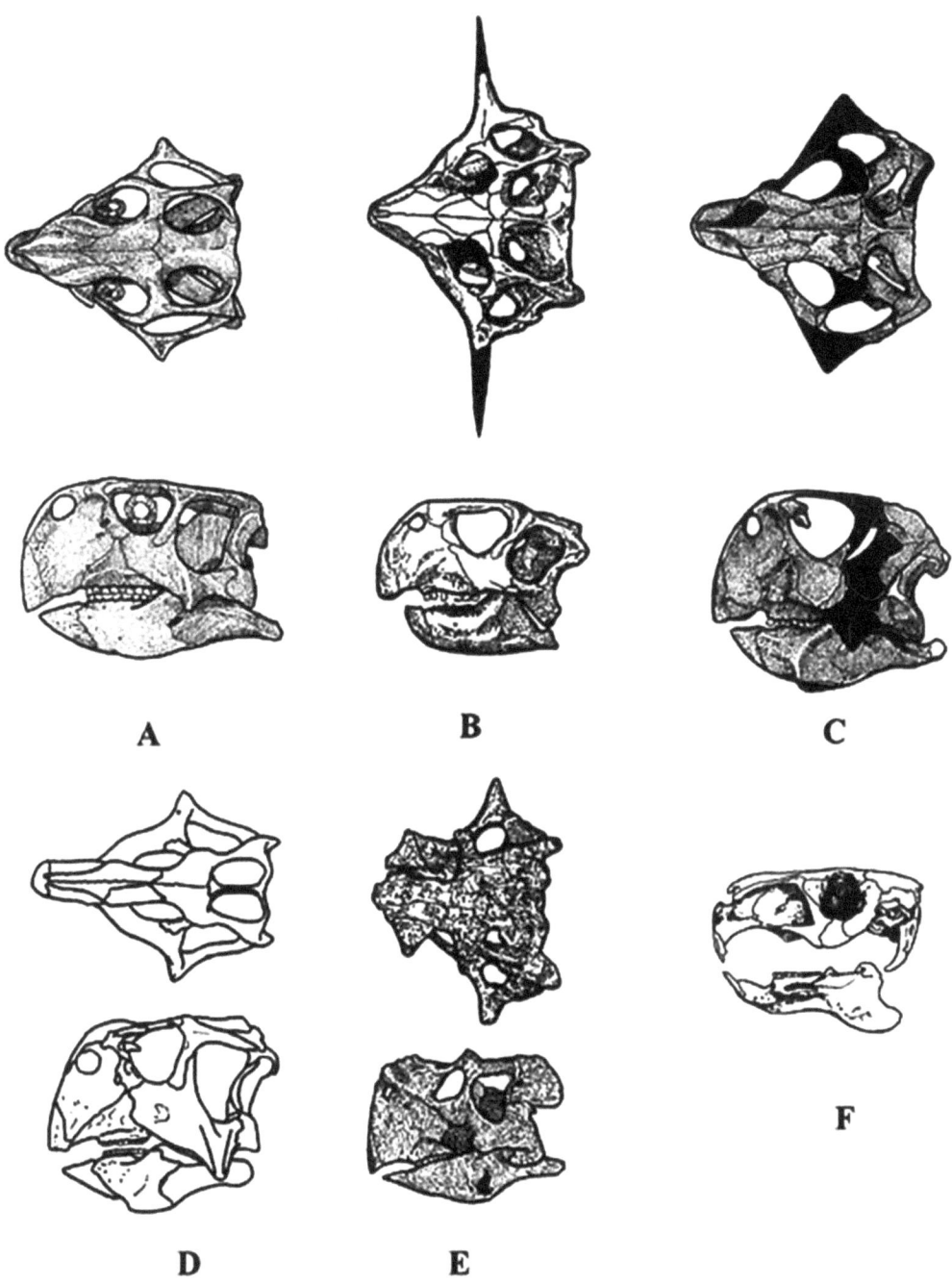

Figure 4). *Psittacosaurus* & capybara skulls. (A) Dorsal and lateral view of *Psittacosaurus mongoliensis* (AMNH 6254) modified from Sereno et al. (1988). (B) *Psittacosaurus sinensis* (IVPP V738), modified from Young (1958). (C) *Psittacosaurus meileyingensis* (IVPP V7705), modified from Sereno et al. (1988). (D) *Psittacosaurus major* (LH PV1), after Sereno et al. 2007. (E) *Psittacosaurus sibiricus* (PM TGU 16/4-20), after Averianov et al. (2006). (F) Skull of capybara (free ware illustration).

141

So, in a nutshell psittacosaurs had several morphologies that led us to believe (as well as Rozhdevensky, Suslov, Currie, and Averianov) may have been semi-aquatic. It had dorso-ventrally flattened toes, a femoral head at a 30°, a femoral head with a cartilages cap which allowed for sprawling hind limbs, dorsally placed naris and orbits, gastroliths used for ballast, front limbs that indicate they had a swimming stroke and would have been a strong swimmer, not to mention the majority of specimens found in an aquatic environment (Figure 7). *Psittacosaurus* may have used several modes of motion to swim; by use of the tail in a similar manner as crocodilians, sprawling its hind legs and using its flattened feet to help propel/kick when swimming, and front leg stroke to help propel and steer.

As I mentioned in the first part I believe *Stenopelix* may be a psittacosaurid, but I'll have to do more research for this possibility.

Are there other ceratopians that have been inferred to as aquatic? As a matter of fact, yes. Tereschenko (2008) published a paper about the adaptive features of protoceratopsids. He believed *Bagaceratops* was mostly aquatic, *Protoceratops* was semi-aquatic and *Udanocertops* facultatively aquatic. Why? Mainly because their tails have tall caudal neural spines, relatively long hamel processes and laterally flattened tail, though this was first discussed by Barsbold (1974). The tail movements was in the horizontal plane and would be useful in swimming than for a terrestrial animal. *Bagaceratops* have extremely tall caudal neural spines. I'm waiting to get more more information on the skeleton of *Bagaceratops* from Tereshnenko at the time of this writing. And yes, there are large lakes and ponds known from the deposits of protoceratopians. He also suggests larger Ceratopsians were semi-aquatic.

There was a poster at the Ceratopsian symposium by Bykowski and Retallack (one was in High School, I think) and their poster was on the possibility of *Triceratops* having a life style like that of a hippopotamus. They basically compared *Triceratops* skeleton to a Hippo, Rhino and Bison. They compared the proximal limb bones, which are more like a hippo. Like *Psittacosaurus* the orbit of Triceratops is high on the skull table, in fact, I've noticed years ago that the orbits are above the braincase. They came to the conclusion the skeleton was more similar to a Hippo, and that the environment that *Triceratops* lived in was wetlands, near large rivers, lakes, and other freshwater deposits. I asked if they'd publisher more on it and they said no. But they weren't the only ones to find Ceratopians skeleton was more like a hippo, Mallon and Holmes redescribed a nearly complete skeleton of an *Anchiceratops* from the Natures Museum in Canada. Langston (1959) hypothesized that *Anchiceratops* elongate face would allow it to breathe while wading in water, its long frill counterbalancing the head. Its short tail, massive body, and stocky limbs "bespeak a sluggish nature that might be expected in animals that enjoyed the relative seclusion and protection of a swampy environment". Semi-aquatic ceratopians goes back even further, Matthew who quoted Hatcher et al, (1907), Matthew, (1915), Gregory and Mook, (1925), Feduccia, (1973) and Russell, (1977), all believed Ceratopians were semi-aquatic. The environment also needs to be looked into and it wasn't as arid as many believe and was more wetland environments. Mallon and Holmes list several characteristics for a semi-aquatic life style; Thickened cortex of limb bones, dorsal placement of nares, dorsal placement of orbits, reduced olfaction, reduced lacrimal, reduced anterior thoracic neural spines, expansive ribcage, large olecranon relative to length of radius, well-developed pectoral muscles, retention of long digital flexor tendons, shortened limbs and I would add, short wide phalanges and unguals.

It may turn out some ceratopains (psittacosaurs, some protoceratopsians and several ceratopains) were more semi-aquatic than is generally perceived. This doesn't mean they lived like a hippo primarily but may have ventured into an aquatic environment on occasion at least.

Bibliography

Averianov, A., Voronkevich, A. V., Leshchinskiy, S. V., and Fayngertz, A. V., 2006. A ceratopsian dinosaur *Psittacosaurus sibiricus* from the Early Cretaceous of West Siberia, Russia and its phylogenetic relationships. Journal of Systematic Palaeontology, 4, 4: 359-395.

Bykowski, R., and Retallack, G., 2007, Was *Triceratops* like a Bison, Rhino or Hippo? Implications for lifestyle and habitat: In: Ceratopsian Symposium, Short Papers, Abstracts, and Programs, complied by Braman, D. R., p. 11-16.

Gregory, W. K., and Mook, C. C., 1925, On *Protoceratops*, a primitive ceratopsian dinosaur from the Lower Cretaceous of Mongolia: American Museum Novitates, n. 156, p. 1-9.

Langston, W. Jr., 1959, *Anchiceratops* from the Oldman Formation of Alberta: Natural History Paper of the National Museum of Canada, n. 3, 11pp

Lu, J., Kobayashi, Y., Lee, Y.-N., and Ji, Q., 2007. A new *Psittacosaurus* (Dinosauria: Ceratopsia) specimen from the Yixian Formation of western Liaoning, China: the first pathological psittacosaurid. Cretaceous Research, 28: 272-276.

Mallon, J. C., and Holmes, R., 2010, Description of a complete and fully articulated chasmosaurine postcranium previously assigned to *Anchiceratops* (Dinosauria: Ceratopsia): In: New Perspectives on Horned Dinosaurs. The Royal Tyrrell Museum Ceratopsian Symposium, edited by Ryan, M. J., Chinnery-Allgeier, B. J., and Eberth, D. A., Indiana University Press, Part Two, p. 189-202.

Matthew, D., 1915. Dinosaurs, with special reference to the American Museum of Natural History. American Museum of Natural History, 129pp.

Mayr, G., Peters, D. S., Plodowski, G., and Vogal, O., 2002. Bristle-like integumentary structures at the tail of the horned dinosaur *Psittacosaurus*. Naturwissenschaften, 89: 361-365.

Sereno, P. C., Chao S., Cheng Z., and Rao C., 1988, *Psittacosaurus meileyingensis* (Ornithischia: Ceratopsia), a new psittacosaur from the Lower Cretaceous of Northeastern China: Journal of Vertebrate Paleontology, v. 8, n. 4, p. 366-377.

Sereno, P. C., Zhao, X., Brown, L., and Tan, L., 2007, New psittacosaurid highlights skull enlargement in horned dinosaurs: Acta Palaeontologica Polonica, v. 58, n. 2, p. 275-284.

Tereschenko, V. S., 2008, Adaptive features of protoceratopoids (Ornithischia: Neoceratopsia): Palaeontological Jounal, v. 42, n. 3, p. 273-286.

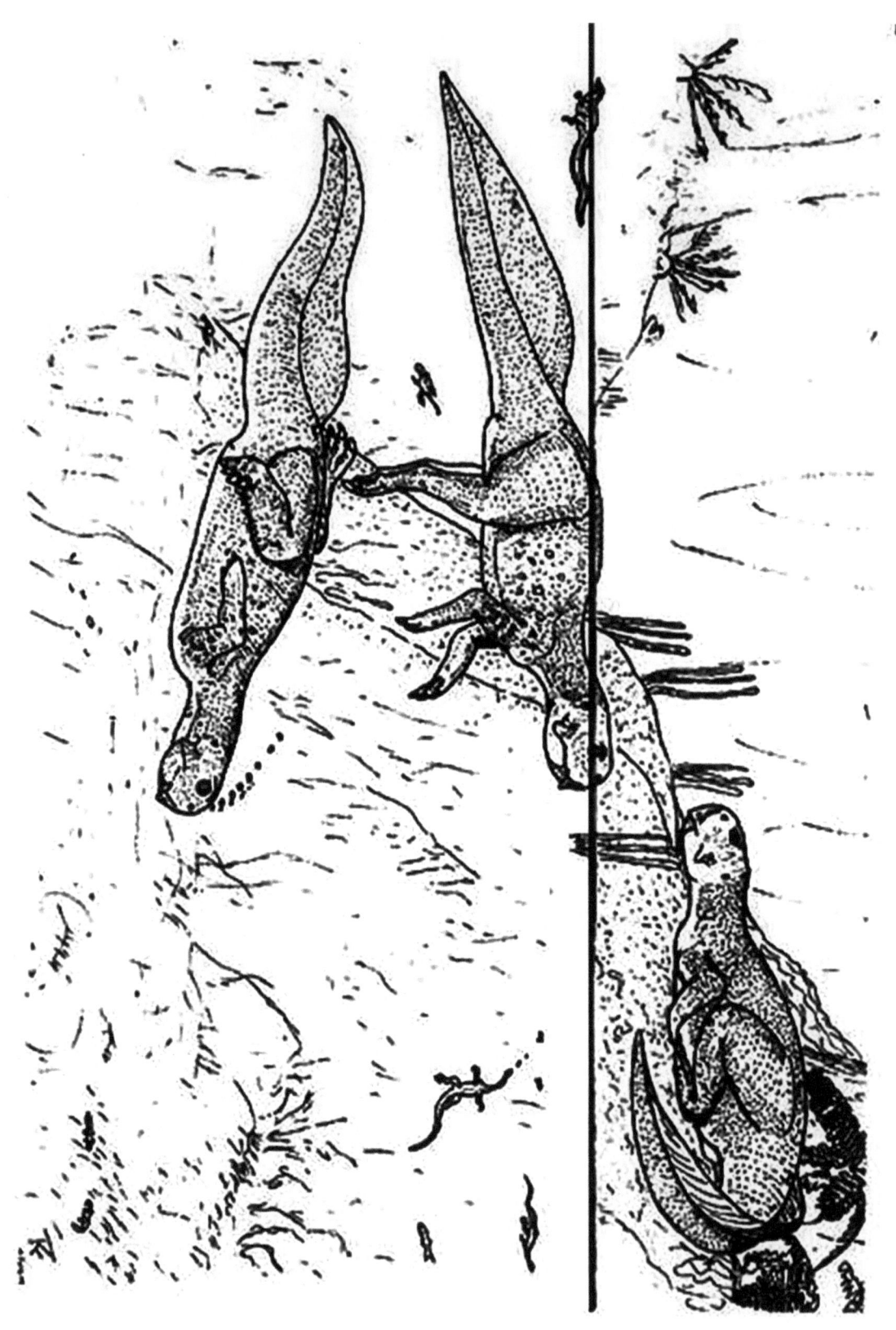

Figure 5). Life reconstruction of aquatic psittacosaurs (This wasn't used in the original article).

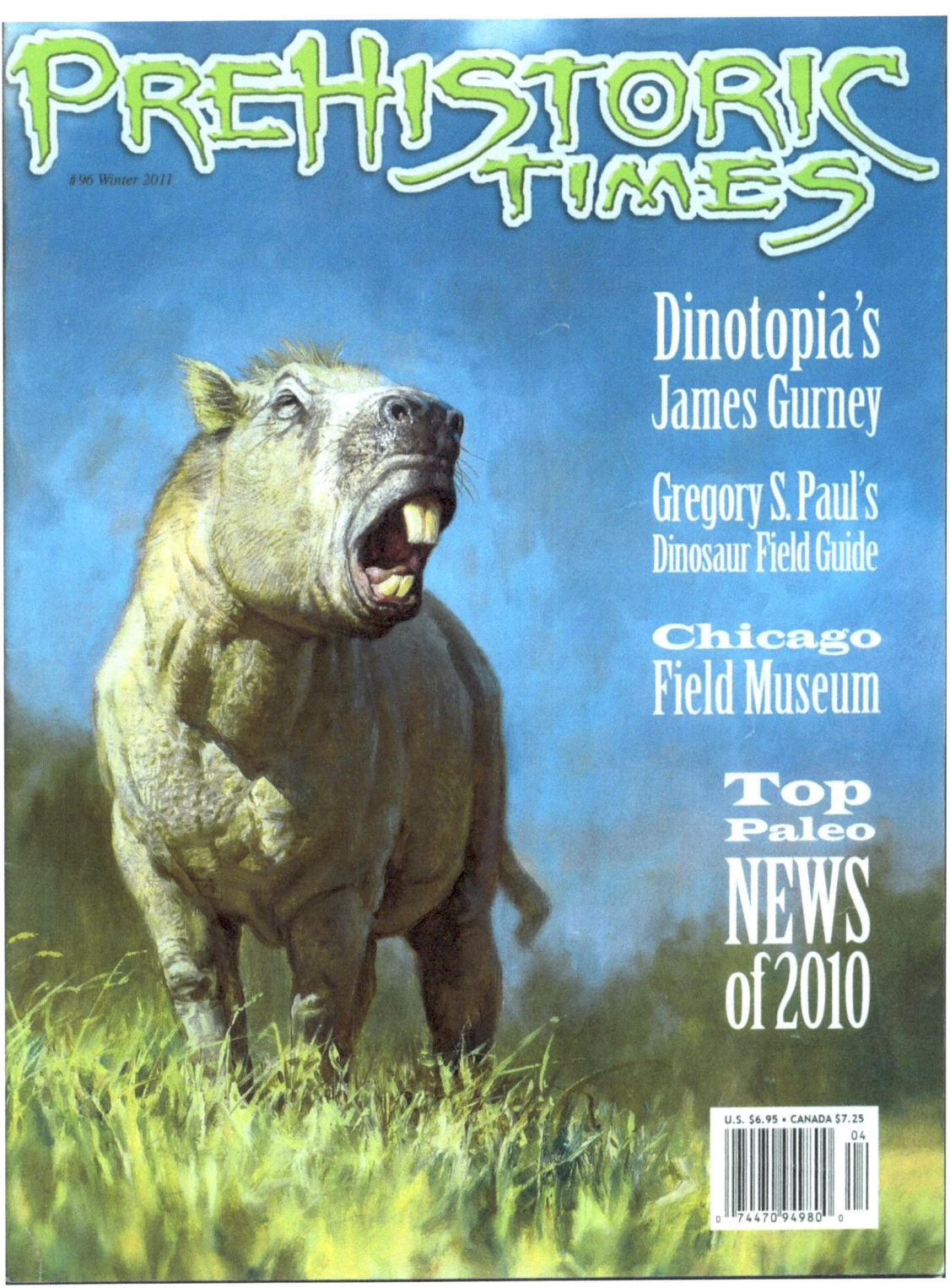

Ford, T. L., 2011, How to Draw Dinosaurs, Aquatic Pittacosaurs (finally) Part two. Prehistoric Times, n. 96, p. 18-19.

Chapter 24

To know the nose, Part 3 (Sauropods)

I first wrote about dinosaur's noses way back in 2002 (PT 51, 52 [How to Draw Dinosaurs volume 2]). I stated that I would write about Sauropods noses latter. Well now after 8 years it is later. I started this series of articles about the research of Dr. Larry Witmer on the noses of extant animals and his inferring that research onto dinosaurs. I wasn't a big fan of Dr. Larry Witmer's theory that dinosaur noses were lower or placed rostrally on the head which is contrary to what was usually understood, but I'm more than happy to change my mind when I'm shown the evidence.

This year's Society of Vertebrate Paleontology Symposium (2011) was held at Philadelphia and was sponsored by the Carnegie Natural History Museum. I was walking around the new dinosaur exhibit at the Carnegie Museum with my friend Kris. We stopped at the prep lab to look in. They have a large glass 'wall' that the public can look in and see what they are working on and to ask questions. The prep lab was open, and they had on display some of the dinosaurs they were working on and a few they had from the collections. On display was the skull of *Diplodocus longus* (CM 11161). As I looked the skull (through the glass) I started to notice the structure of the skull. That led me to the possibility that Dr. Witmer was right.

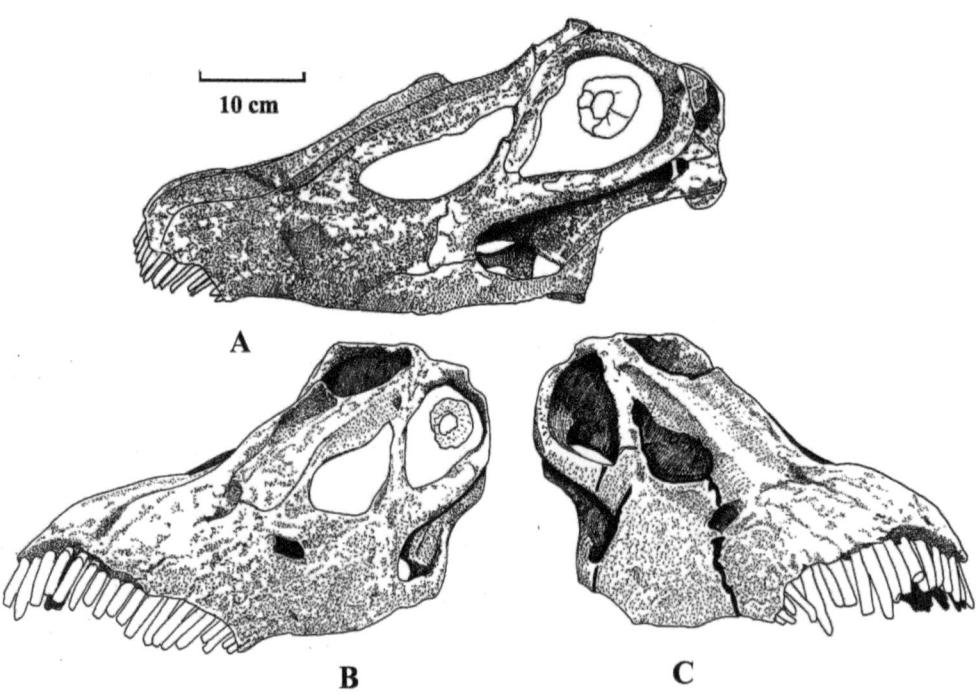

Figure 1) Skull of *Diplodocus longus* CM 11161; A) Side view; B) Left view; C) Right view.

First, we need to understand the skull of *Diplodocus* (CM 11161) (Figure 1). The nasals in *Diplodocus* are just in front of and above the orbits. They are fairly small and flat. They form a square or rectangular plate on the top of the skull. The premaxillaries are very long bones that are about half the skull length. The narial opening itself is about one fourth the length of the skull and the back of the skull takes up the last fourth of the skull length. The nasal opening isn't located on the top of the skull as is typically believed but is in front of the orbit and faces more forward than dorsal. The back of the skull is at 25 % angle to the horizontal with the apex at the anterior edge of the nasals and at the mid section of the orbit. From the anterior end of the nasals and at the mid section of the orbits the dorsal region of the skull angles

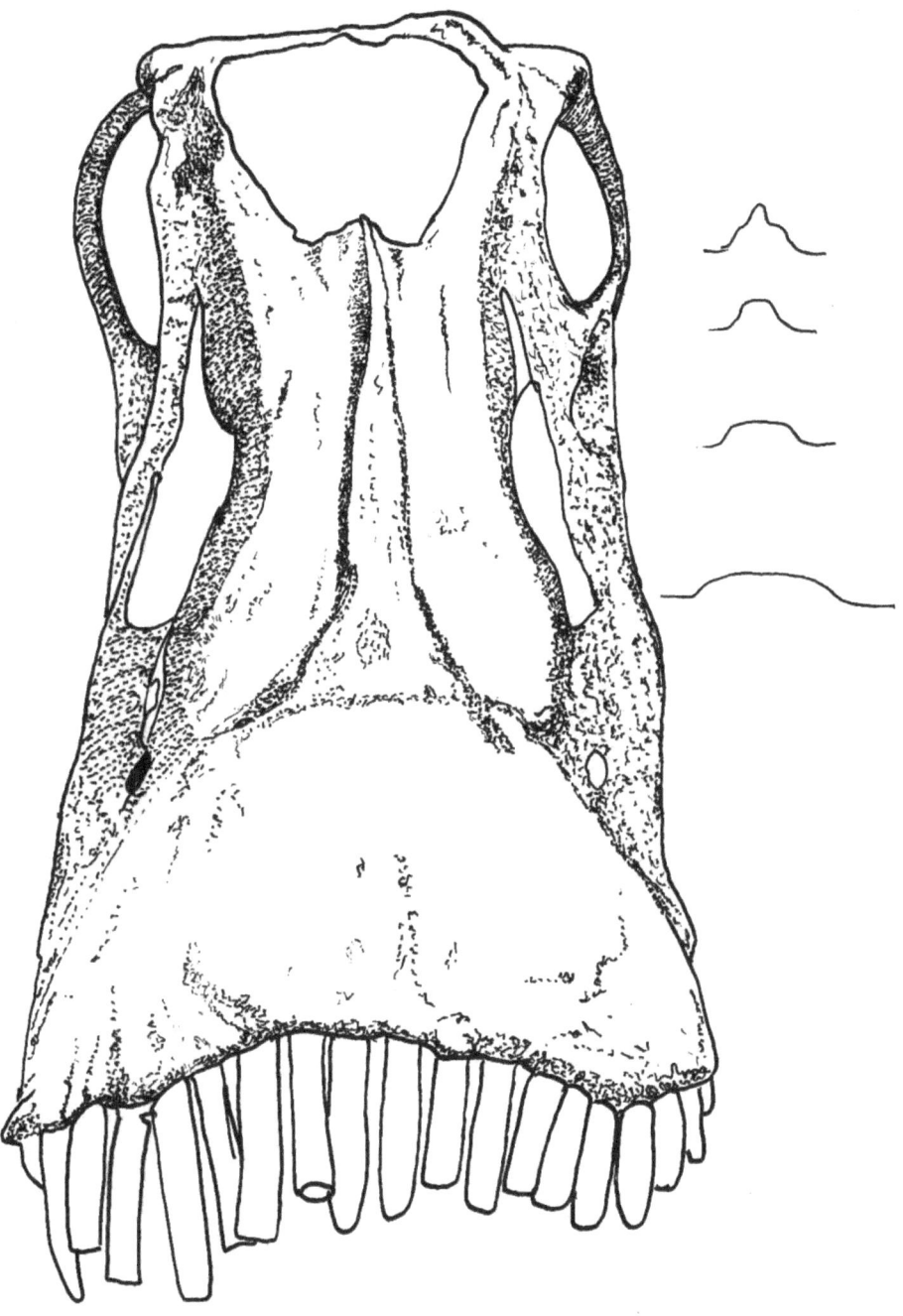

Figure 2) Skull of *Diplodocus longus* CM 11161. Front view and a schematic view of the ridge and shelf.

in the opposite direction at about 30 % anterior (or rostrally using Dr. Witmer's terminology). I'll explain why that is important shortly. The posterior edge of the paired premaxilla where it ends at the narial opening has a pronounced medial, laterally compressed ridge (Figure 2). Anterior/rostrally this ridge fans out slightly and becomes a short raised 'shelf' to about where the premaxillae levels off toward the end of the mouth, or at one fourth from the tip of the snout. The paired premaxillae dorsally from this 'shelf' to where it meets the maxilla is fairly broad and flat and shows a slight indention in the premaxilla/maxilla.

As I studied the skull (through the glass) I noticed that on the outside of the ridge and 'shelf' the dorsal area of the premaxillae has a flattened area (dorsally) with a very shallow 'groove' where the premaxillae level off and the shallow 'groove' ends and expands laterally in a broad area. This got me

147

thinking about what the 'groove' was for and the soft tissue structure of that area or to put it another way, that indicated to me something was there and it wasn't just a flat skull. The premaxillae ridge may have acted like as septum that extended into the narial opening by a cartilage septum (which has been speculated before). The soft tissue of the 'nose' itself would have extended from the narial opening and followed the groove in the premaxilla to the leveled off area of the premaxilla forming a large 'fleshy' narial opening. This area may have had a cartilaginous structure giving the fleshy nares structure. This agrees well with Witmer's work.

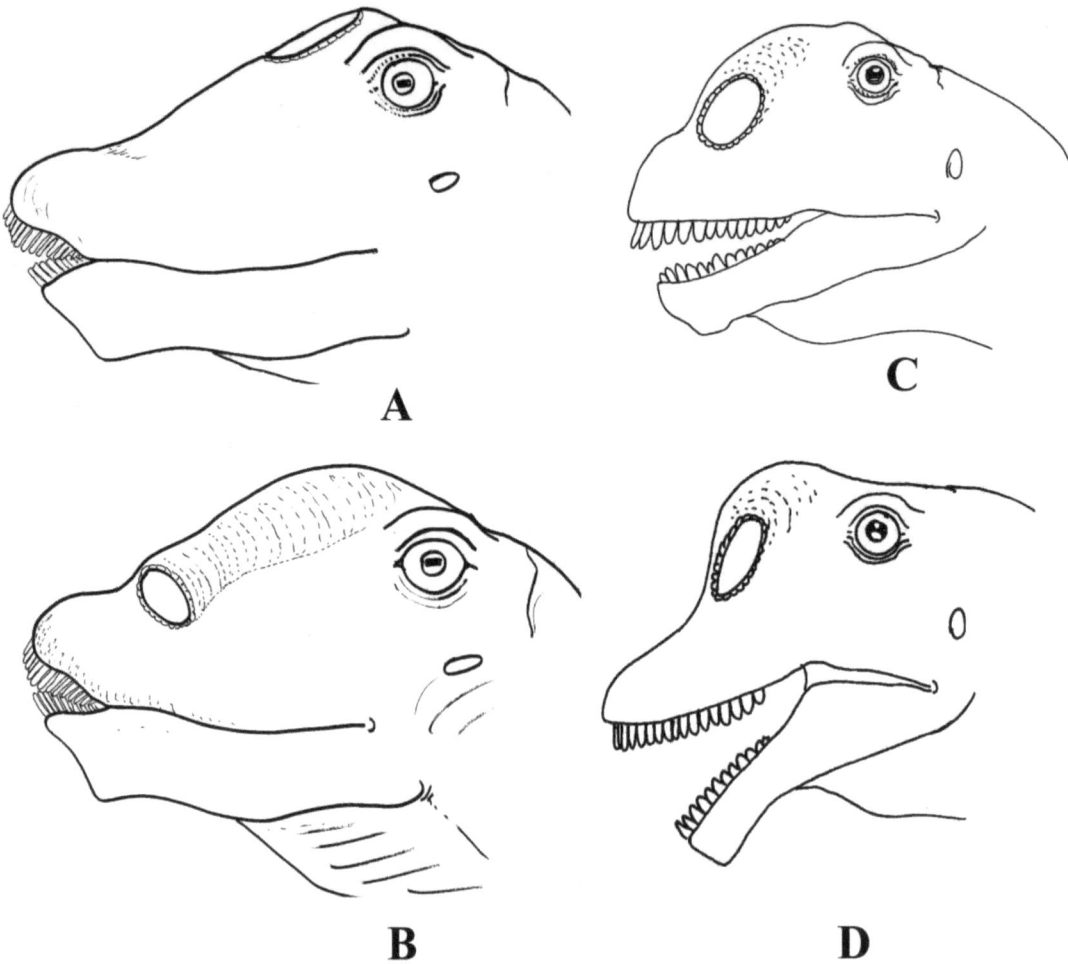

Figure 3) Head of Sauropods; A, B) Diplodocus *Diplodocus longus* CM 11161; A) Old interpretation of head and nose; B) New interpretation of nose and head; C) *Camarasaurus* head showing new interpretation of nose and head; D) *Brachiosaurus* sp from North America showing new interpretation of nose and head.

How big was the 'fleshy' narial opening? Could the 'fleshy' narial opening have been larger to account for the air flow needed to feel the lungs? While I was listening to a talk about breathing and lung size in alligators I started to think about sauropods. How big would the 'fleshy' opening have had to be for the animal not only to breath, but also to fill the pneumatic bones. The air would have to flow down the windpipe of its long neck, to its huge lungs and the pneumatic bones. I believe the 'fleshy' nose would have had to have been larger than Dr. Witmer depicted. If you take the angle of back of the skull (25%) and extend it forward to half the length of the narial opening, then follow the angle of the front of the premaxillae (30%) down to the leveled off area would make the 'fleshy' nose fairly large. This would make the sauropod look very strange. What about carmarasaurs and brachiosaurs? They have large narial opening. I also believe the fleshy' nose would have been near the front of the narial opening and would have been larger than is usually believed (Figure 3). At the reception at the Carnegie I flagged down Dr.

Witmer and pointed out to him CM 11161 and that I now agreed with him. He said that that was the skull he used in his research.

As I was told by a friend of mine at the SVP and reading other's comments (and also in Dr. Witmer's paper) about position of the nose in animals, they like to smell what they are eating. So, the 'fleshy' nose near the front of the head is a good thing. The rostrally placed naris would have also helped in thermoregulation and selective in temperature regulation of the brain.

And finally, I would like to make a few comments about my last articles. Also, at Carnegie there was a cast of the psittacosaur with bristles on the tail. What I found interesting was that as the caudal vertebrae dipped downward but the skin outline continued horizontally. The 'flesh' skin impression of the tail is thicker at the end of the preserved tail. Could the 'skin' continued and made the tip of the tail "thicker"? Or is it just an artifact of preservation? Also, where the back of the bent legs are is a vertical straight skin impression. This would have made the hind legs 'thicker'. These may have also helped in swimming. (Figure 4).

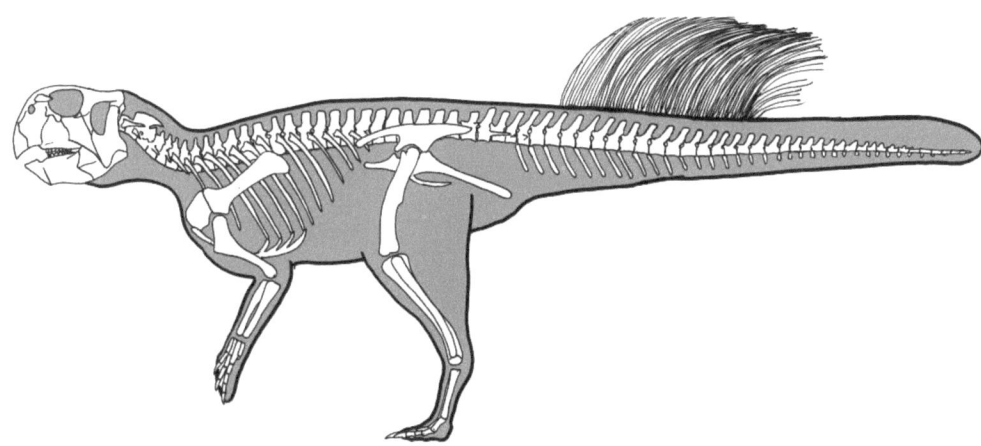

Figure 4) New interpretation of the skeleton and tail of *Psittacosaurus* based on the cast of the *Psittacosaurus* with the bristle tail at the Carnegie.

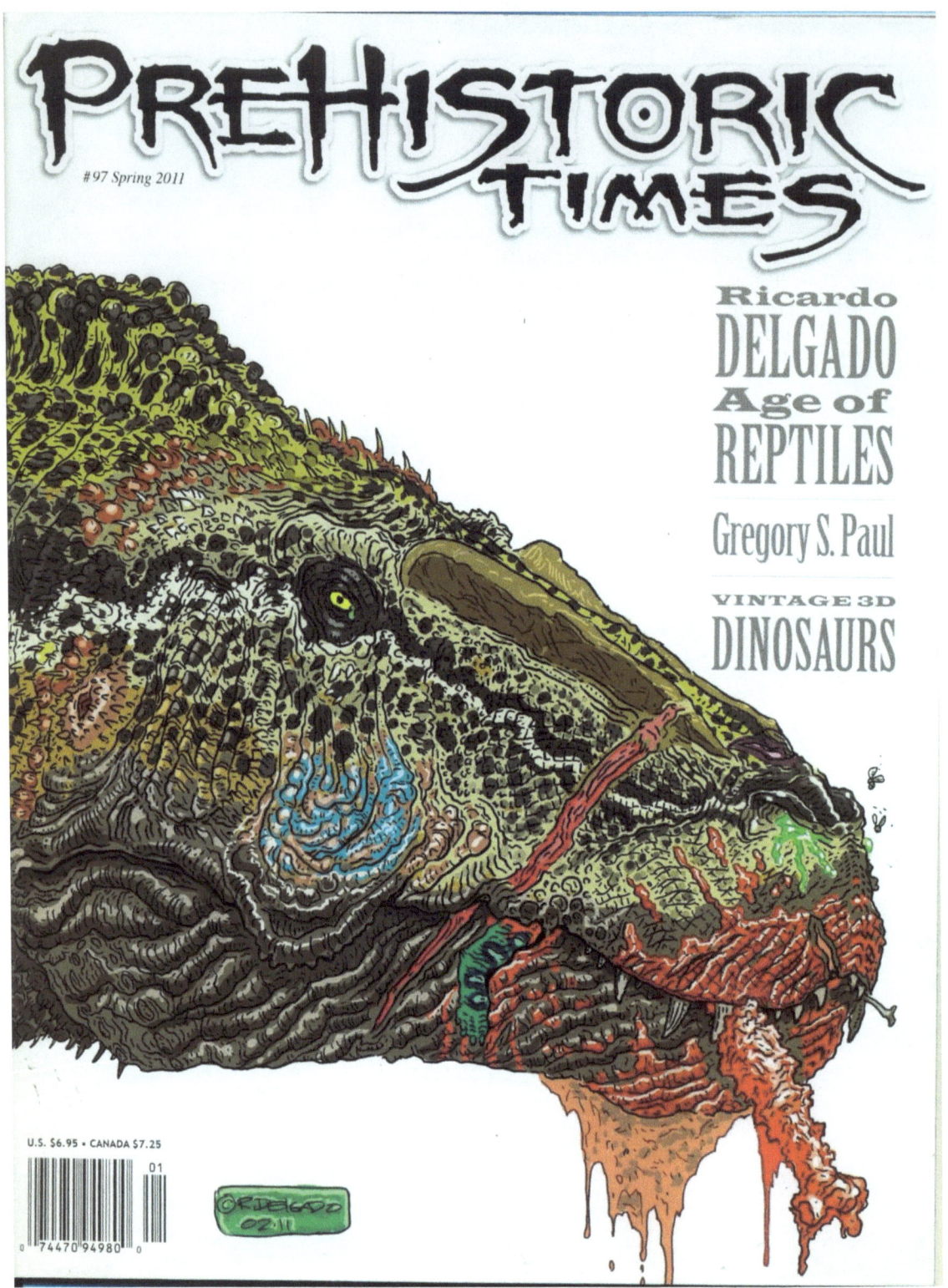

Ford, T. L., 2011, How to Draw Dinosaurs, And now, the end of dinosaurs. Prehistoric Times, n. 97, p. 18-19.

Chapter 25

And Now, The End Of Dinosaurs

Well, actually I'll be discussing the muscles of the tail of dinosaurs. I will be talking about Ornithopods and Theropods this issue and next issue I'll be taking on Sauropods and whether or not sauropods could stand in a tripodal stance to eat and then ankylosaurs and stegosaurs.

Recent articles about breathing in dinosaurs (Carrier & Farmer, 2000) and the pelvic and tail muscles in *Tyrannosaurus rex* (Persons IV and Currie, 2011) inspired me to write this article. Persons dissected several lizards and the spectacled caiman to determine the insertion and attachment of the tail muscles and it is from this research that they inferred that information onto *T. rex*. I infer the information from these articles to other dinosaurs.

I'll be talking about the following muscles (and will follow Persons IV and Currie, 2011 descriptions); *m. ilio-ischiocaudalis* (= *ischio-caudalis*), *m. longissmus* (= *ilio caudalis*?), *m. caudoformalis brevis*, *m caudi-femoralis longus*, and the c*audotruncu*s (Carrier & Farmer, 2000). The attachment/insertion areas for the tail muscles are the posterior section of the ilia, ischia, pubes, the chevrons/hamel arches, centra, and transverse processes. These muscles help in balance and walking.

It is interesting to note that the majority of articles on the pelvic muscles leave out the *caudotruncus* as do many artists. The *caudotruncus* muscle helped in breathing (Carrier & Farmer, 2000). I won't be getting into breathing in dinosaurs but will take it up in a latter issue. For our purpose we are interested in appearance not function. The *caudotruncus* inserts on the ventral side of the pubis, slides/glides along the ventral tip of the ischium and attaches to the ventral side of the first three chevrons. There are two sides of the *caudotruncus* because it has to allow room for the cloaca. The *caudotruncus* is also small in ornithopods. (Figure 1).

The *m. ilio-ischocaudalis* has two major subdivisions; *m. iliocaudalis* (ilium) and *m. ischiocaudalis* (ischium). The *m. ilio-ischiocaudalis* is relatively thin anteriorly where it attaches dorsally to the lateral tips and ventral surfaces of the transverse process. It raps around the *m. caudoformalis* and attaches ventrally to the ventral tips of the hemal spines and to its bilaterally symmetric muscular doppelganger that raps around from the other side. Posteriorly the *m. ilio-ischiocaudali*s increases in relative thickness where the *m caudoformalis* diminishes. After the disappearance of the *m. caudofemoralis,* the *m. ilio-ischiocaudalis* inserts onto the full lateral surface of the centrum and the hemal spines to about the fifth caudal, except in hadrosaurs that have a longer ischium and may have attached to the 10th or 11th caudal. In ornithopods this muscle is huge because the ischium is long and in hadrosaurs it is very long. In the majority of theropods the *m. ilio-ischocaudalis* is large but in dromaeosaurs/troodontids etc it is very small. (Figure 2).

The *m. caudi-formalis* anteriormost insertion is the fourth trochanter and the transverse processes cap its dorsal most limit and the ventral most margins of the hemal arches are its ventral extent. It has two 'sections'; the *m. caudoformalis brevis* and the *m. caudofemoralis longus*. The *m. caudoformalis brevis* fills the brevis fossa and may also insert at the 4th trochanter and across the anteriormost caudal vertebrae and the *m. caudofemoralis longus*. It attaches to the 4th trochanter. The *m. caudofemoralis longus* also inserts medially onto the lateral faces of the caudal vertebrae and inserts the lateral faces of the hemal spines, though this does vary. It attaches onto the 4th trochanter. The further away the transverse processes are from the pelvis the longer the muscle is. In *Tyrannosaurus rex* the muscle attaches to the 14th caudal vertebra, in *Allosaurus* around the 19th, in *Ceratosaurus* is around the 29th and in dromaeosaurs it attaches to the 8th caudal vertebra. In ornithischians *Hypsilophodon* attaches to the 14th, *Camptosaurus* around the 12th, the 10th in *Iguanodon* and around the 15th in *Gryposaurus*. This would mean *Ceratosaurus* had the most muscular tail in the listed dinosaurs and dromaeosaurs with the least. (Figure 3).

Anteriorly the *m. longissimus* inserts onto the full dorsal surfaces of the transverse processes and the later faces of the neural arches, dorsal area of the ilium and along the dorsal neural arches of the caudals. Posteriorly after the transverse processes have terminated, the lateral surfaces of the neural arches are the only osteological insertion. The longer the tail the longer the muscle. And the *m. spinalis* inserts on the dorsal tips of the neural arches and attaches to the dorsal posterior end of the ilium. They run the full length of the tail. (Figure 4).

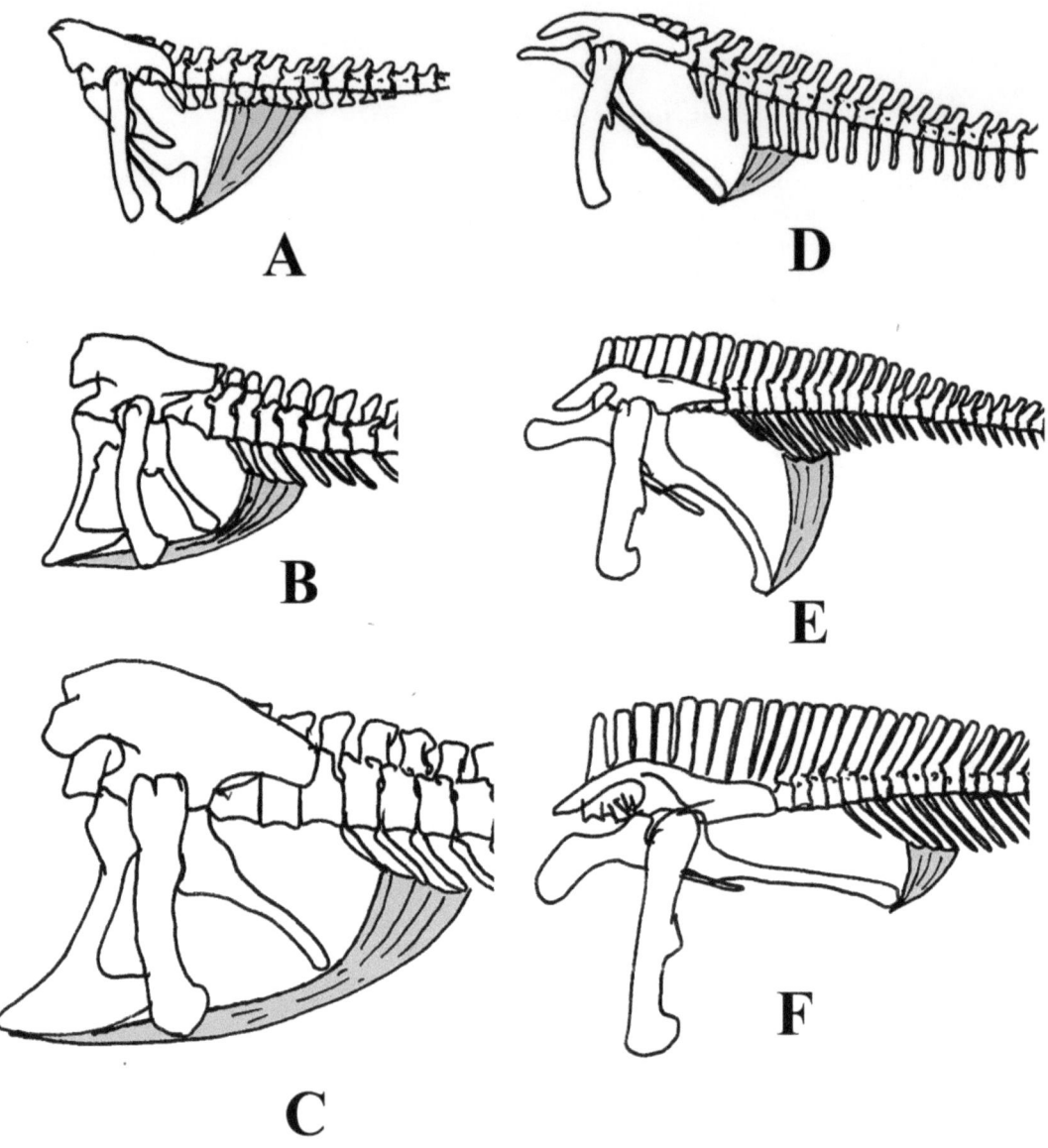

Figure 1) *Caudotruncus* muscle in Grey.

Persons IV and Currie research shows that contrariety to common belief the cross section the *m. caudofemoralis longissimus* was elliptical in shape and not flattened. This would mean the tail had a more oval shape.

What this means is the tail in dinosaurs? As Persons IV and Currie point out there has been a trend to depict non-avian theropods with a "flattened tail". Their research shows that the tail would have looked more like a crocodile without its armor. At the transition point the tail would have been laterally compressed also at the tip of the tail. But in between that the tail would have been as broad or broader laterally as they were dorsoventrally. Along with the *caudotruncus* the pelvic area would have been thicker. (Figure 5).

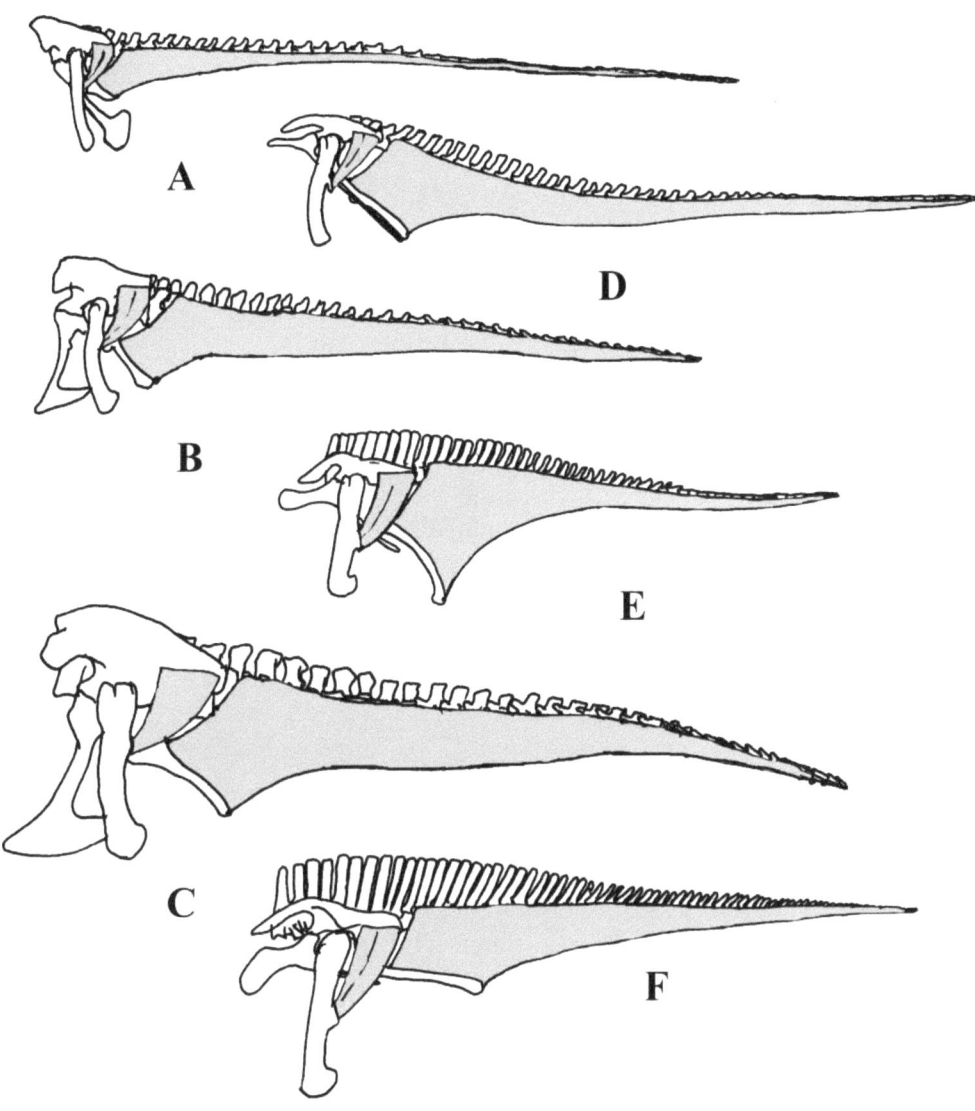

Figure 2) *M. caudoformalis brevis* which inserts on the 4th trochanter and attaches to the posterior ventral side of the ilium, and the *m. ilio-ischocaudalis*.

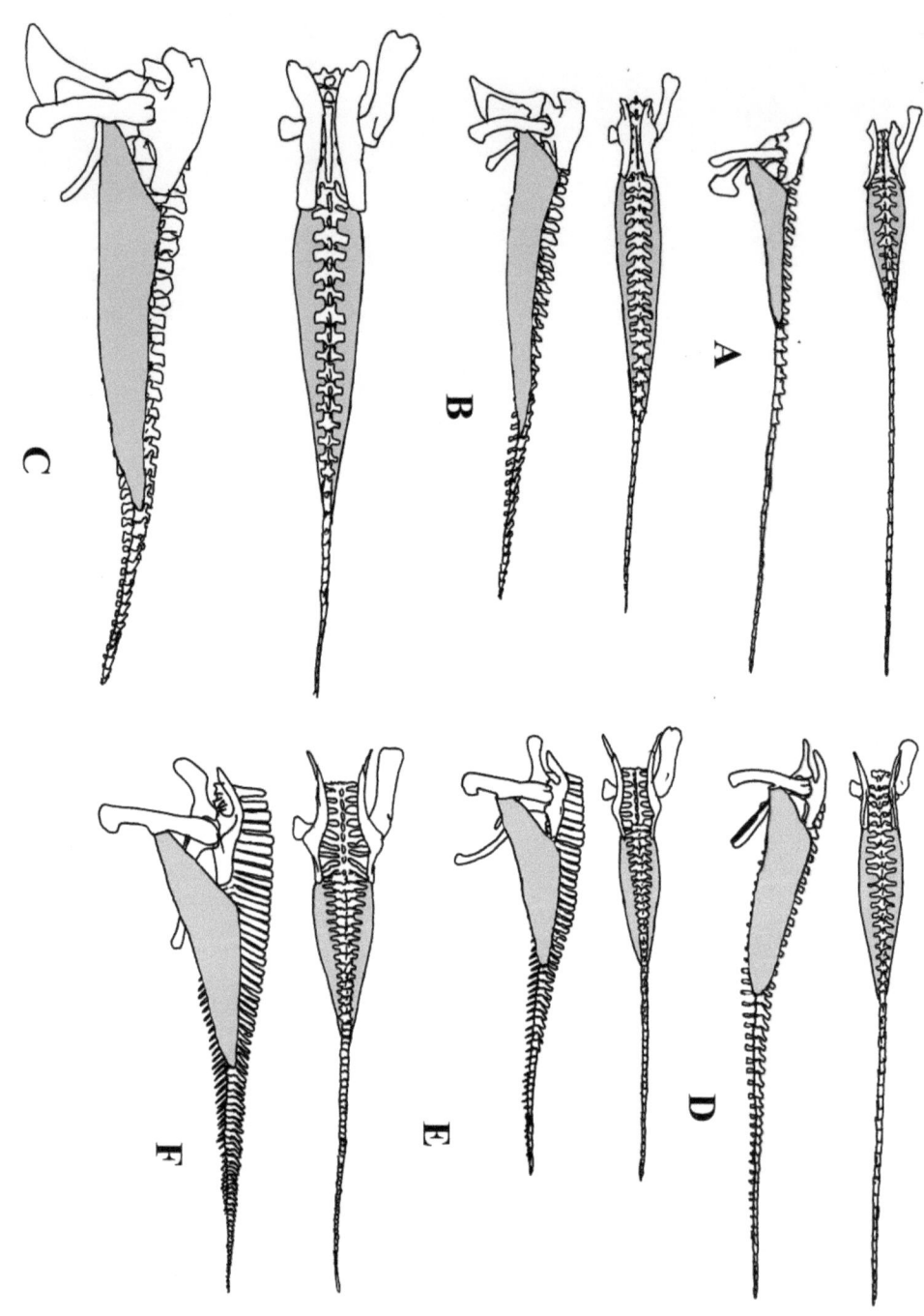

Figure 3) The *m. caudofemoralis longus*.

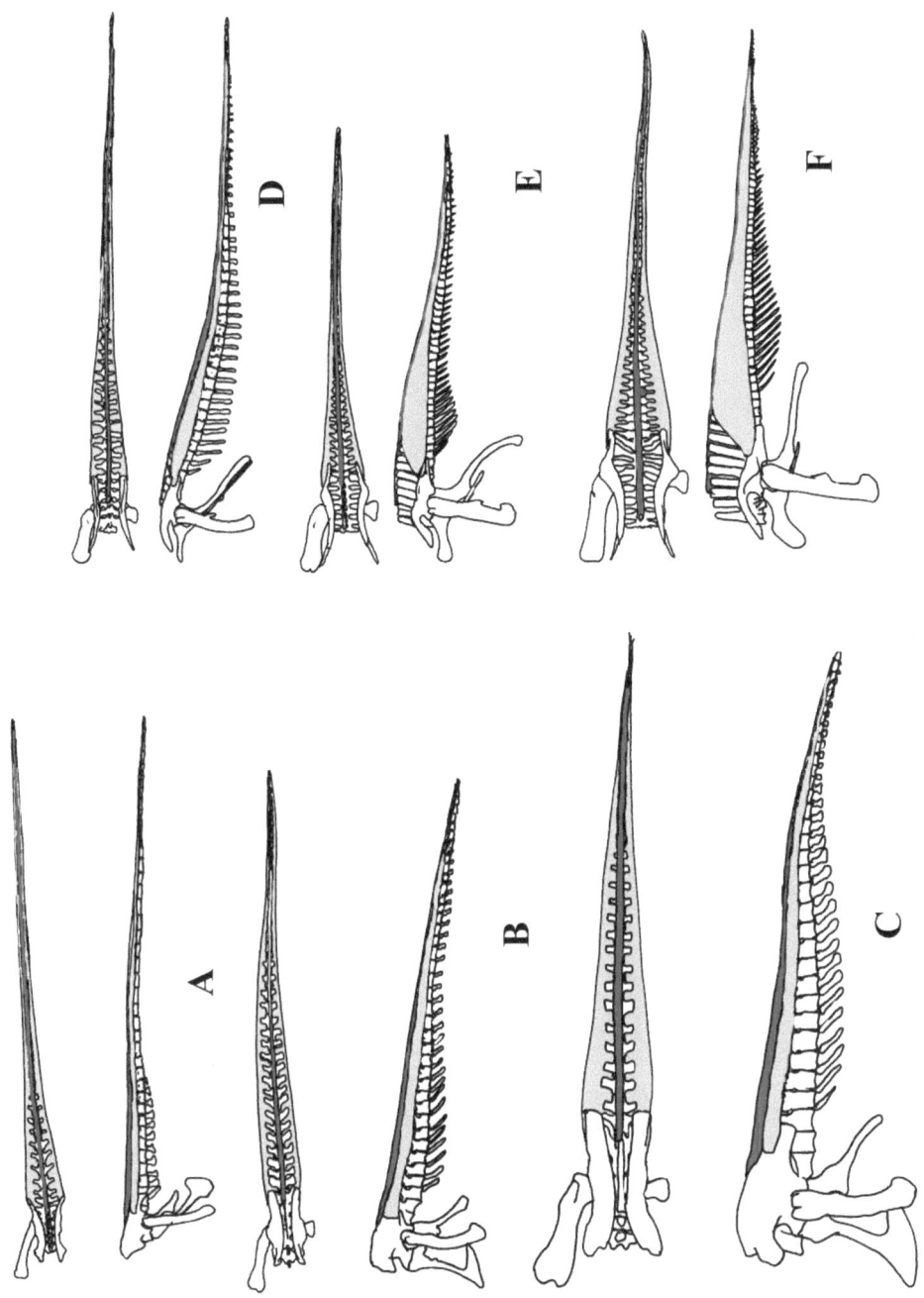

Figure 4) The *m. longissimus* and *m. spinalis*.

Bibliography

Carrier, D. R., and Farmer, C. G., 2000, The integration of ventilation and locomotion in archosaurs: American Zoologist, v. 40, p. 87-100.

Paul, Gregory S., 1996. The complete Illustrated Guide to Dinosaur Skeletons. Gakken Mook:

Persons IV, W. S., and Currie, P. J., 2011, The tail of *Tyrannosaurus*: reassessing the size and locomotive imporance of the *M. caudofemoralis* in non-avian theropods: The Anatomical Record, v. 294, p. 119-131.

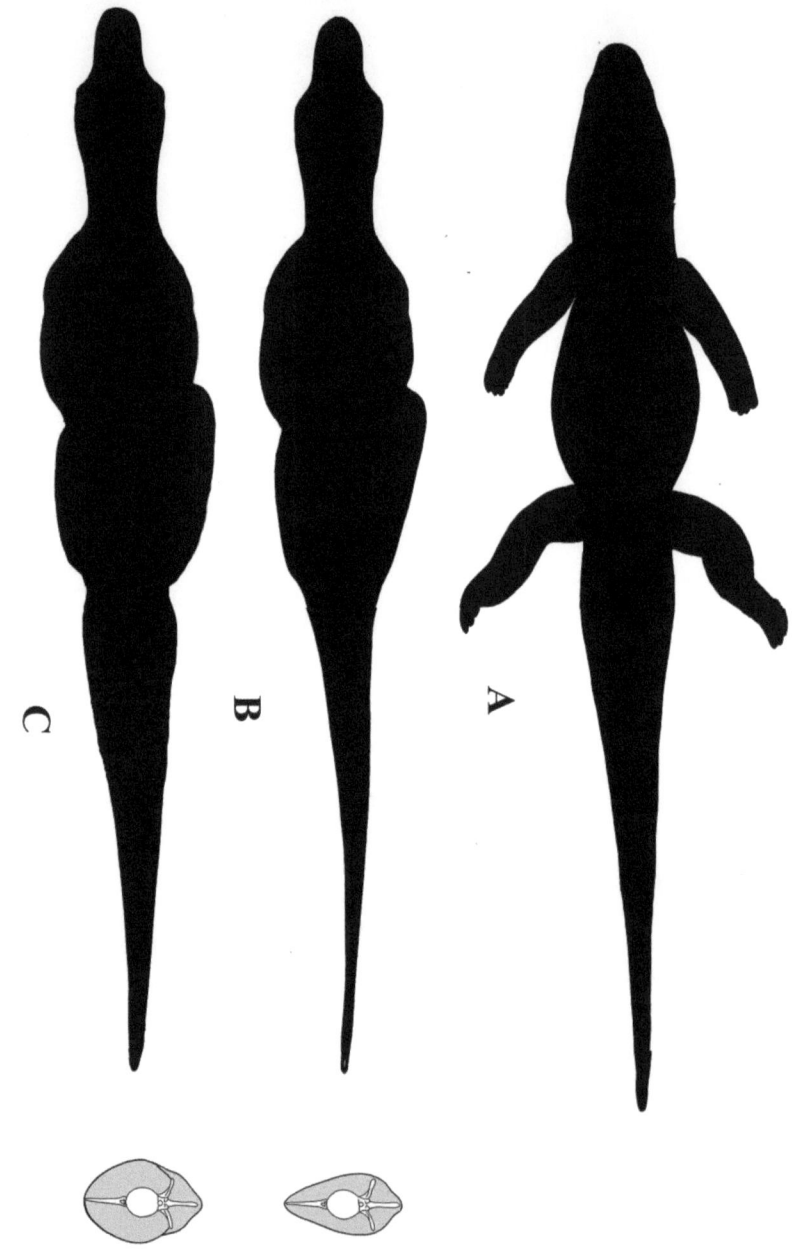

Figure 5) Silhouette illustrations of an Alligator A) and *Tyrannosaurus rex* B, C). B) Showing the old version of a 'thin' tail with cross-section of *Allosaurus* tail modified from Madsen, 1976; C) New interpretation of a thicker tail from Persons & Currie and a modified tail from Madsen, 1976.

I encourage the readers to also purchase my other books; How to Draw Dinosaurs volume 1, Volume 2, Generick Dinosaur Skull-a-Day Volume 1, Volume 2, my co-authored coloring books with Mike Frederick, What Color were Dinosaurs? What Color were Prehistoric Sharks and Rays? And my novel, Dinosaur Isle.

Generic Dinosaur Skull-a-Day Calendar
Volume 1
By Tracy Lee Ford

Generic Dinosaur Skull-a-Day Calendar
Volume 2
By Tracy Lee Ford

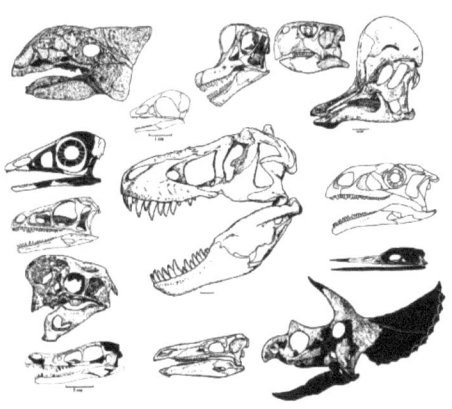
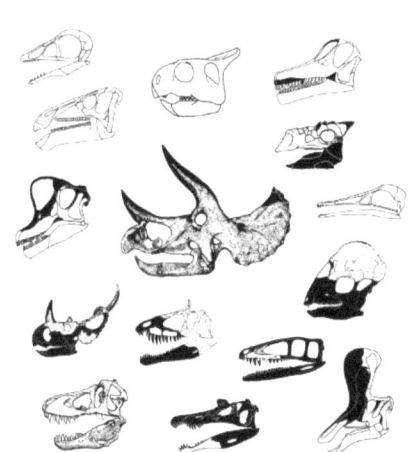

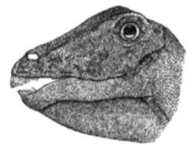

How to Draw Dinosaurs

Volume 1

By Tracy Lee Ford

How to Draw Dinosaurs

Volume 2

By Tracy Lee Ford

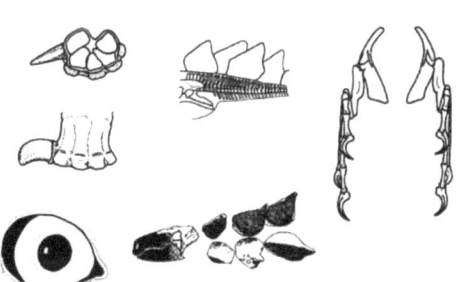
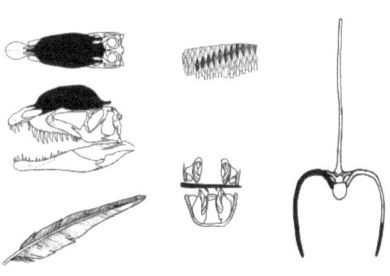

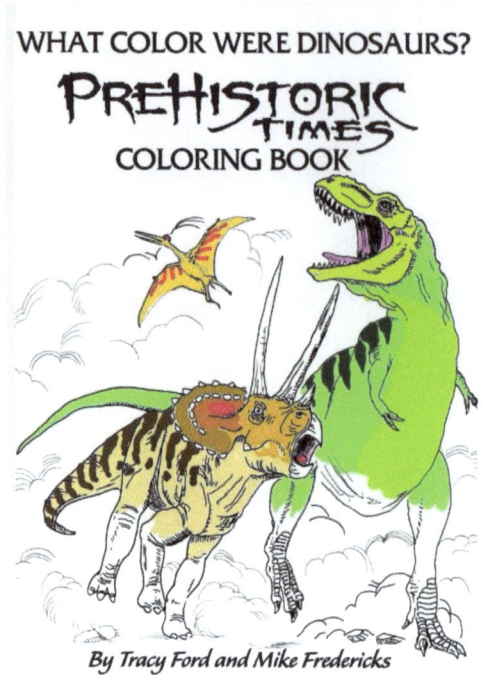
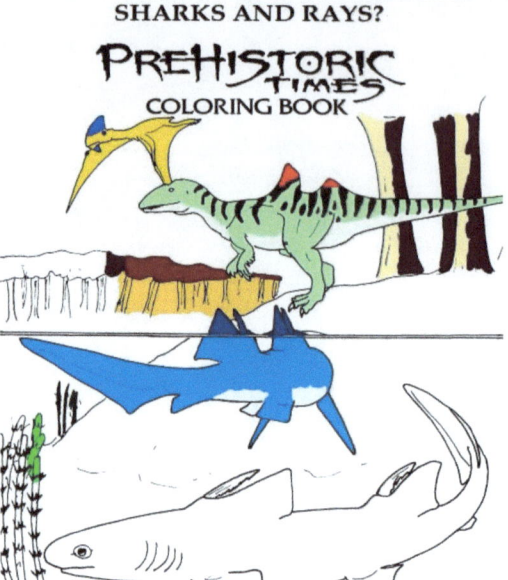

www.ingramcontent.com/pod-product-compliance
Lightning Source LLC
Chambersburg PA
CBHW051147220526

45473CB00003B/681